D0031727

RELIABILITY AND YIELD PROBLEMS OF WIRE BONDING IN MICROELECTRONICS

The Application of Materials and Interface Science

by

George G. Harman

National Institute of Standards and Technology

International Society for Hybrid Microelectronics

The International Society for Hybrid Microelectronics (ISHM) is a non-profit technical society dedicated to the advancement of microelectronics. The Society's prime objectives are to provide a forum for the dissemination of knowledge within the field of microelectronics and to serve as a common denominator for the diverse engineering disciplines on which microelectronics is based.

ISHM encourages the exchange of information among the complementary technologies of ceramics, thin and thick films, semiconductor packaging, discrete semiconductor devices, and monolithic circuits. Microelectronics has developed into a distinct field of activities embracing materials, design, processing techniques and equipment, and fabrication and applications engineerings. ISHM's technical meetings and publications reflect the full range of these engineering specialties.

ISHM supports more than 30 local chapters and a growing number of student chapters throughout the United States and Canada. ISHM cooperates with affiliated chapters in Europe, the Middle East, and the Far East. An elected Executive Council composed of officers and regional directors determine the policies of the Society. A professional staff, headquartered in metropolitan Washington, D.C., in Reston, Virginia, administers ISHM's programs.

ISHM is a thriving vigorous technical society, immensely proud of its distinguished reputation and totally committed to the service of its dedicated membership and the flourishing microelectronics industry.

ISBN 0-930815-25-4

Printed in the United States of America

Additional copies of this publication can be ordered from ISHM, P.O. Box 2698, Reston, VA 22090-2698

CONTENTS

FOREWORD

The International Society for Hybrid Microelectronics (ISHM) is proud to offer this text, "Reliability and Yield Problems of Wire Bonding in Microelectronics: The Application of Materials and Interface Science." Since its founding in 1967, ISHM has been committed to providing current information in microelectronic technology. Although an earlier monograph on microjoining processes was published by ISHM in 1983[1], this book is the most extensive compilation on current wire bonding issues to appear in print.

In the late 1950s, wire bonding techniques were developed to provide interconnections for the first integrated circuits. Thermocompression and ultrasonic bonding were the first methods employed for interconnecting complex solid state devices. Subsequently, thermosonic bonding incorporated elements of these earlier methods. Wire bonds are used to interconnect semiconductor chips to the exterior contacts of electronic packages. In hybrid circuits, wire bonds interconnect chips-to-chips, chips-to-passive components, and to exterior contacts of the electronic package. Wire bonds provide highly reliable contacts to electronic devices. In a similar manner, wire bonds are employed to interconnect passive and active devices in chip-on-board structures.

[1]D.D. Zimmerman and D.H. Lewin, "The Fundamentals of Microjoining Processes," ISHM , Reston, pp. 1-80 (1983).

This book describes the conditions for making reliable wire bonds with a high yield by describing all potential sources of failures, from the final stages of wafer processing, through handling, bonding, testing and screening. Sources of contamination are identified that adversely affect the reliability of wire bonds. In addition, the degrading effects of temperature, temperature cycling, and mechanical forces such as ultrasonic cleaning are described. Bonding machine setup parameters also play a critical role. In addition, the severity of the above problems may depend on the ambient atmosphere, the metallurgy of the wire, and/or the morphology of the bonding pad metallization. Wafer sawing and die attach can also adversely affect bond quality.

Basic concepts of bonding methods, wire metallurgy and aging, and cleaning techniques (UV and/or ozone, solvent, plasma, and burnishing) are described. Classical plague failure, its metallurgy, and the effect of corrosion and impurities are extensively treated. All bond testing methods are described and compared. Problems with electroplating, various metal systems, and machines and setup are described. Thermal and ultrasonic effects on wire fatigue are discussed. Mechanical problems as cratering, cracks in wedge bonds, and the effect of acceleration and vibration are extensively given.

With this book, George G. Harman gives a very detailed description of up-to-date reliability and yield problems of wire bonding in microelectronics. Mr. Harman has combined his world-class expertise in wire bonding with experiences from teaching on this technology to produce a much needed text on this subject.

ISHM is grateful to the author and the National Institute of Standards and Technology for this opportunity to disseminate knowledge on wire bonding technology that is vital for assuring the manufacture of quality microelectronics.

Bernard S. Aronson
Director of Technical Services
International Society for Hybrid Microelectronics

INTRODUCTION

Twenty years ago, wire bonds caused a large proportion — sometimes as high as one-third — of all semiconductor device failures. However, the number of recognized failure mechanisms at that time was quite limited. Typically, they were cited as "purple plague," underbonding, overbonding, and nonspecified contamination. There was an effort to employ aluminum ultrasonic bonding (versus gold thermocompression) to avoid "purple plague," but this merely switched the susceptible bond from the chip to the package, which was generally gold-plated. It did, however, lower the bonding temperature from 300^{+} °C to 25 °C which did help in some cases.

At the present time (1989), dozens of chemical, metallurgical, and mechanical failure mechanisms have been identified. Part of these new mechanisms were discovered because of greatly improved analytical methods and equipment (e.g., Auger and SIMS analysis), part because of the trillions of bonds made (and millions to billions failed), and part because of the changing technology (e.g., new metallurgies, plastic encapsulation). A study of recent wire bond failure papers indicates that the discovery of new failure mechanisms has slowed, although the rediscovery of old ones, or variations of them, has continued unabated. Thus, it was felt that this is an appropriate time to review the known failure modes and mechanisms of wire bonds, categorize them, and where possible, explain and/or give solutions to them.

Since failures are generally revealed or screened by testing, the bond pull and shear tests are described. The review will discuss mechanical, metallurgical, chemical, and miscellaneous failure mechanisms, although some will overlap and will be placed in the most appropriate sections.

This work will be primarily concerned with failure mechanisms and yield problems originating from chip-to-package wire bonds, although other wire bonds, such as crossovers and PC board bonds, may be included as appropriate. Flip chip, TAB, beam leads, lead-frame bonding, etc., will be included where wire bond-type interfaces and failure mechanisms have been observed. But the book is ***applicable*** to many of these cases. For example, in the case of B-TAB (bumps on the tape) it is usual to plate the copper lead with gold. Then either thermocompression (TC) mass bonding or thermosonic (TS) point bonding is used for welding. The metallurgical interfaces, as well as the bonding methods, are identical to gold-ball bonding to a chip. Thus, discussions of chip cleaning, gold plating, and gold-aluminum intermetallics are as applicable to this format TAB (or its variation — ball-bond-bumped TAB) as to wire bonding. In general, the metallurgy and the interface reliability of any TAB inner lead process using point bonding is addressed.

This book has been written to serve as a text to accompany courses taught at the ISHM International Microelectronics Symposium, the University of Arizona, and others. It is written at a practical level and is intended for use by production line engineers in solving, or avoiding, bonding problems. However, enough detail and/or references are included for failure analysis personnel or others who are interested in doing research on the subject. Areas where more research is needed are clearly indicated. It is hoped that workers will study and fill in these gaps in the near future rather than spend time adding a fifth decimal to the well characterized problems.

It is assumed that all readers have a basic knowledge of wire bonding, device packaging and/or hybrid circuit assembly technology. Terms in general usage in these fields will not be defined, but others that are less well known (such as fracture toughness) will be described. The book is divided into relatively independent chapters and, with its index, can be used to look up specific problems.

Manuscript organization. Each section of this book is self-contained with its own references and numbering system. In some cases, this results in the same reference appearing at the end of two or more sections. This was thought to be most convenient for the reader. The sections are coupled together by referring from one to another for greater detail or explanation where appropriate.

Manuscript style. The style is that of a technical paper, as is prescribed by NIST rather than the informal style used by the author in his presentations and courses on this subject.

Units. The choice was reluctantly made to use both SI and English units since significant parts of the American semiconductor and hybrid community still use the latter. However, many figures were reproduced directly from technical papers and their units may be in either system alone. Mixed units are frequently used in English language publications. The units of force are probably the most confusing. Grams-force (gf) is most often used in this text. The reader can convert 1 gf = 9.8 millinewtons. For any remaining unconverted units it should be noted that 1 mil = 0.001 in = 25.4 μm.

DISCLAIMER

Certain commercial equipment, instruments, or materials are identified in this book in order to adequately specify the experimental procedure. Such identification does not imply recommendation or endorsement by NIST, nor does it imply that the materials or equipment identified are necessarily the best available for the purpose.

ACKNOWLEDGMENTS

This book could not have been written without the editorial, organizational, and typing work of Mrs. Josephine Gonzalez. The changes and corrections must have seemed to approach infinity. Without her help, the author's thoughts, organization, as well as his office, would be forever in a chaotic mess. The help of Mrs. Gonzalez is immensely appreciated. Additional editorial work and direction by Mrs. Jane Walters was also indispensable.

Various sections of the book have benefited by the technical reviews and/or contributing comments of Dr. John H. Smith, Dr. William J. Boettinger, Dr. Donald B. Novotny, and Dr. Raymond S. Turgel of the National Institute of Standards and Technology; Dr. Peter Douglas and Dr. D. Gareth Davies of American Fine Wire; and Dr. Ronald Thiel of General Dynamics. Many technical authors have graciously contributed original copies of their photographs to be incorporated in this book. Their contributions are referenced in each figure. ISHM personnel, including Bernie Aronson and Denise Hudson, have been enormously cooperative and helpful.

Finally, I wish to acknowledge my division management's support and dedication of resources for the book. Without these resources, the work would not have been attempted.

CHAPTER 1
WIRE BOND TESTING

1.1 INTRODUCTION

Although this book is primarily concerned with the yield and reliability of wire bonds, the normal method for evaluating these problems involves some form of testing. The most common method for evaluating wire bonds is the pull test, both destructive and nondestructive. Details of these tests will be addressed below. While the pull test is valid for wedge bonds, it is necessary to use a shear test to evaluate ball bonds adequately. Therefore, the ball shear test is included also. The theory and applications of the above tests are fully understood, and the tests have become standard ASTM test methods. The pull tests are currently called out in military specifications, and the ball shear test is expected to be included in the future. The "pluck" test or "lift" test, as well as the thermal stress test, will also be briefly discussed because of their implications for bond reliability.

1.2 BOND PULL TEST

The wire-bond pull test is the most universally accepted method used for controlling the quality of the wire bonding operation. This test has

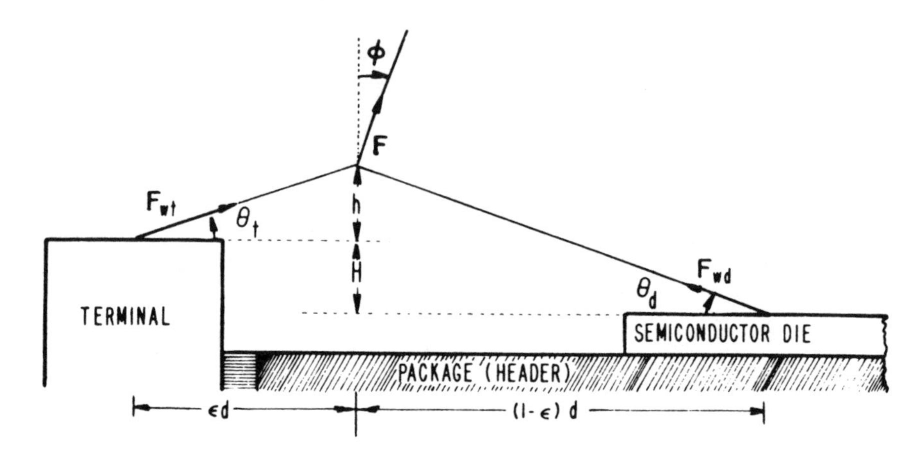

Figure T-1. Geometric variables for wire-bond pull test in plane of bond loop as used in equations (T-1) through (T-4) [T-4].

been used to evaluate the strength of wire bonds since its introduction for semiconductor devices in the 1960s.

The objective of this section is to examine the variables of the bond pull test, both theoretically and experimentally, and from them determine the most likely sources of problems and errors inherent in the test as it is typically performed.

1.2.1 Variables of the Bond Pull Test

Numerous papers have been written on the subject of the wire-bond pull test. Derivations of the equations, standard test methods, and their validation in round-robin tests have variously been given by Polcari [T-1], Schafft [T-2], Albers [T-3], and Harman and Cannon [T-4].

In order to understand the intricacies of the wire-bond pull test, it is necessary to consider the geometrical configuration as well as the several equations that define the resolution of forces. The geometrical configuration defining the variables is given in Figure T-1. The force in ***each wire***, at break, with a specified pull force at the hook, F, is:

$$F_{wt} = F\left[\frac{\{h^2 + \varepsilon^2 d^2\}^{1/2}\{(1-\varepsilon)\cos\phi + \frac{(h+H)}{d}\sin\phi\}}{h + \varepsilon H}\right] \qquad (\mathrm{T}-1)$$

$$F_{wd} = F\left[\frac{\left\{1 + \frac{(1-\varepsilon)^2 d^2}{(H+h)^2}\right\}^{1/2}(h+H)\left\{\varepsilon\cos\phi - \frac{h}{d}\sin\phi\right\}}{h + \varepsilon H}\right]. \qquad (\mathrm{T}-2)$$

If the hook is applied near a bond, the vertical peel component of the force can approach the force of a 90 deg pull test (MIL STD 883, Method 2011). The equations are therefore relevant when testing TAB leads near the chip.

If both bonds are on the same level (H = 0), the loop is pulled in the center (ϵ = 0.5), and then the more familiar equation is obtained:

$$F_{wt} = F_{wd} = \frac{F}{2}\sqrt{1 + \left(\frac{d}{2h}\right)^2} = \frac{F}{2sin\ \theta} \qquad (T-3)$$

where $\theta_t = \theta_d = \theta$. Note that, in general, for bonds of a given strength, larger values of h/d will result in higher pull-force values.

Equivalent equations using angles θ_t, θ_d and F are:

$$F_{wt} = \frac{F}{sin\theta_t + cos\ \theta_t\ tan\ \theta_d} \qquad (T-4)$$

$$F_{wd} = \frac{F}{sin\theta_d + cos\ \theta_d\ tan\ \theta_t}. \qquad (T-5)$$

A calculated plot of pull force at wire rupture is given for typical two-level semiconductor device bond configurations in Figure T-2 pulled in the center of the loop. With everything else being equal, it is apparent that the higher the loop height, the higher the bond pull force will be.

The placement of the hook (ϵ in Figure T-1) or the pull angle Ø will significantly affect the distribution of forces at the bonds. One can choose an ϵ or Ø value that will give equal forces on each bond, and it will result in a more equal test of both bonds. This is practical with the present generation of automated pull testers. However, manual pull test operators would be significantly slowed by such a procedure. In addition, most specifications (such as MIL STD 883 and ASTM F459) require that the hook be placed in the center between the bonds.[1] So this will probably remain the preferred hook placement position for normal testing. Whenever the hook is moved close to one bond or the other, a higher proportion of the wire force is in the vertical or peel direction. If bonds have a tendency to peel, then a significantly lower pull force will result as shown in Figure T-3. Details of this and other special pull test pitfalls (e.g., pulling out of

[1]TAB inner leads are often pulled as close to the chip as possible. See Harman, G. G., Acoustic-Emission-Monitored Tests for TAB Inner Lead Bond Quality, IEEE Transactions on Components, Hybrids, and Manufacturing Technology, Vol. CHMT-5, pp. 445-453, December 1982.

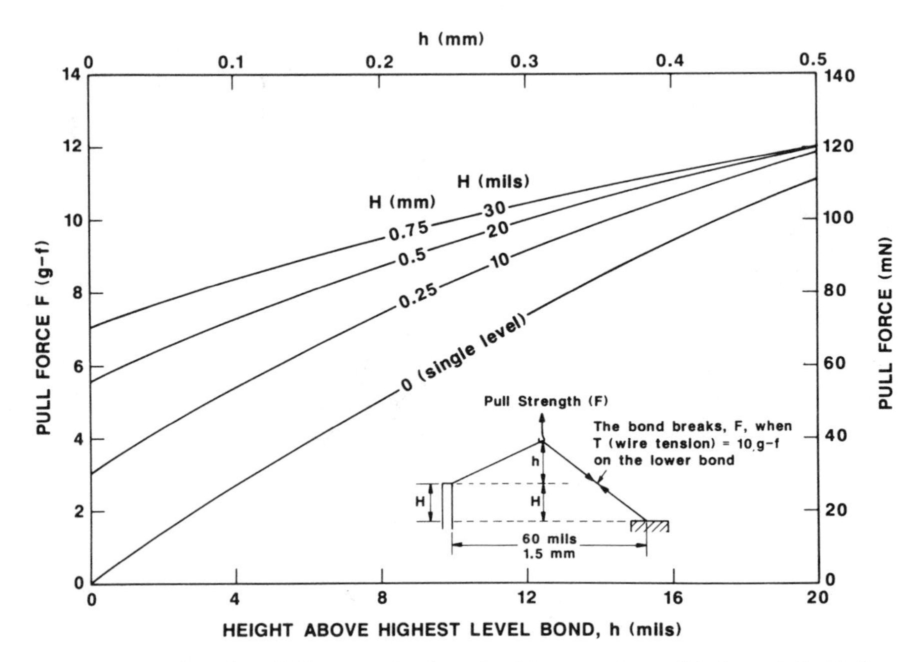

Figure T-2. Bond pull force calculated with equations (T-1) and (T-2) for various loop heights and terminal (post) heights, pulled in the center of the loop.

the plane of the bond, the effect of one weak and one strong bond, etc.) are given in reference [T-4].

1.2.2 Effect of Metallurgy and Bonding Processes on the Bond Pull Force

In a production-line environment where speed is essential, pull test operators seldom ascertain that the hook is at the exact center of the bond loop. The hook will tend to slip toward the highest point of the loop. This point is determined by the type of bonding machine or by the package, if it has a very high or low bond pad. Hook slippage[2] can lead to peel-mode failures as described previously. If both bonds are well made, however, the method of bonding will generally dominate the results.

Gold ball bonds (thermocompression or thermosonic) are bonded with a capillary-type tool. The wire rises straight up from the center of the ball, bends with a high peak almost over the ball bond, and progresses linearly

[2]Some pull testers have stiff hooks that eliminate slippage and pull vertically regardless of the shape of the loop. These are preferable.

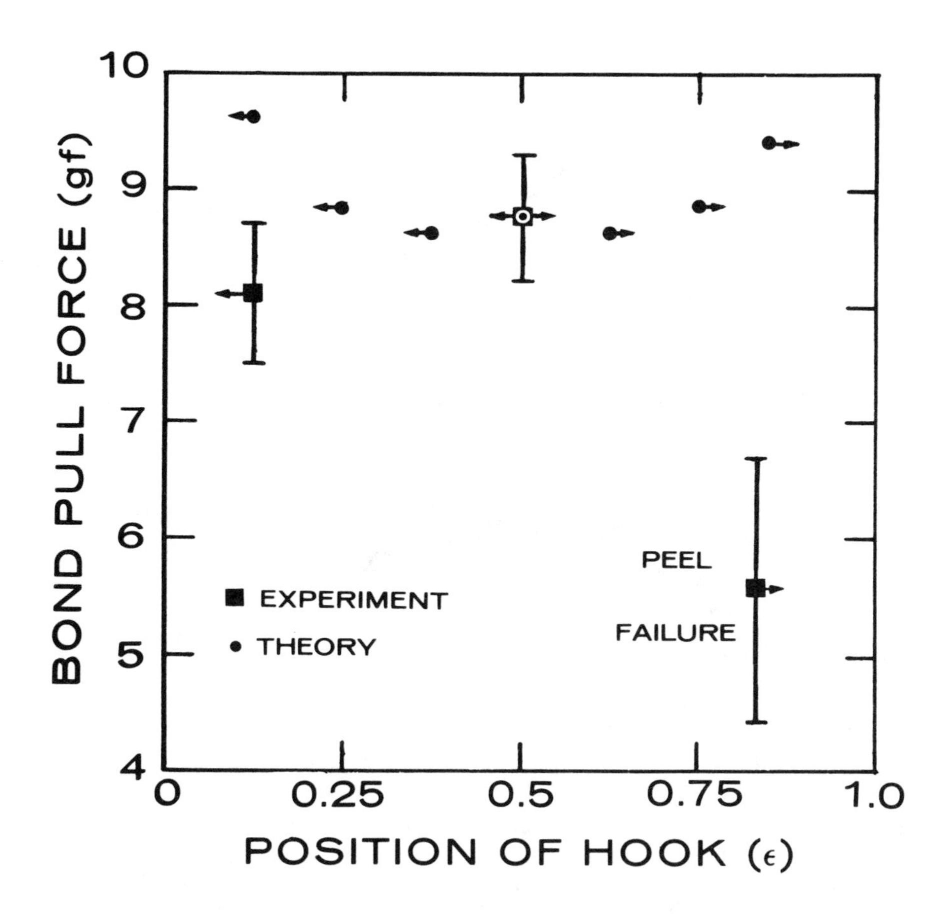

Figure T-3. Measured and calculated pull force as a function of hook position for single-level, 25-μm (1-mil) diameter ultrasonic aluminum bond pairs having first and second bonds with equal breaking strengths, d = 1.5 mm (60 mil) and h = 0.35 mm (14 mil). First bond is located at $\epsilon = 0$ and second bond at $\epsilon = 1.0$. Experimental data: ←▪, Each point is mean of 25 to 30 bonds pulled at indicated hook position. Error bars represent ±1 standard deviation of mean. Failure occurred at bond indicated by arrows. Centerposition breaks ($\epsilon = 0.5$, bond angles - 25 deg) were all tensile failures, 60% of which occured at heel of first bond. All second bonds lifted (peel failure) when pulled in position $\epsilon = 0.85$ (second bond angle = 60 deg). Theoretical prediction: ←•, each point is calculated from equations (T-1) and (T-2) assuming that both first and second bonds are of equal strength and all failures are by tensile-mode breaks. Arrow points to position of bond that would break; centerposition breaks are evenly divided between two bonds (after Harman and Cannon [T-4]).

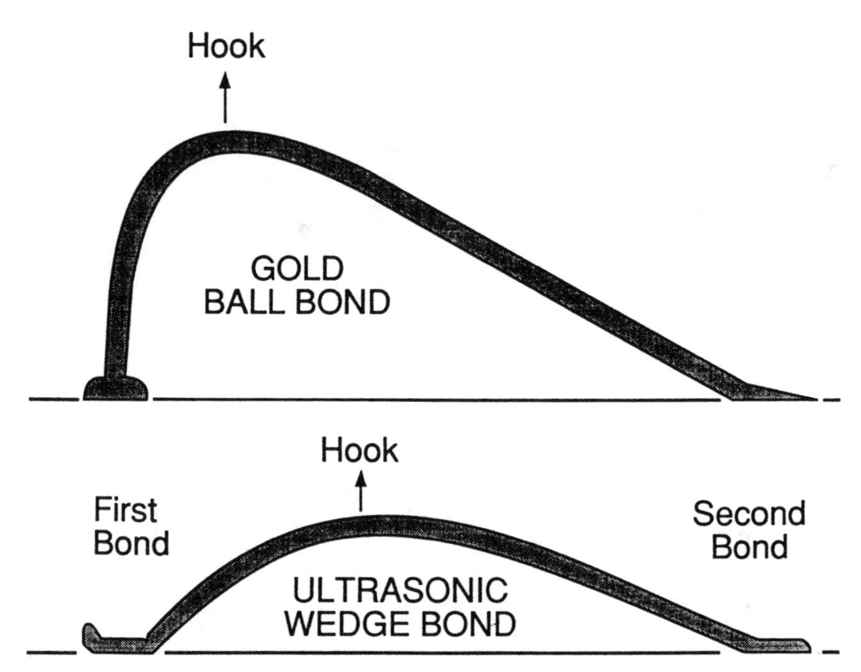

Figure T-4. Typical geometrical configuration and position of pulling hook for gold ball bond (top) and ultrasonic wedge bond (bottom). Pulling hook tends to slip toward peak of loop.

downward towards the second bond, a wedge bond (see Figure T-4). If the pulling hook rises to the peak, most of the force is applied directly to the ball bond which, because of its large area, is generally stronger than the wire. (Ball bond peeling or tearing is not affected by this off-center hook placement as it is for wedge bonds.) Typically, the wire breaks in the recrystallization zone immediately above the ball when applying a force very close to the breaking load of the original wire. The second bond (the wedge or crescent bond) is usually weaker than the ball bond. However, when the hook is located at the peak of the loop (above the ball), relatively little force is applied to the wedge bond, and it seldom breaks.

Thus, only the stronger bond (the ball) is tested (see the section on shear testing for testing the ball bond). For ultrasonic wedge bonds, Figure T-4, the story is just the opposite. The wire rises from the edge of the first bond (in this case, the weaker bond), peaks somewhat before the center of the relatively low loop, and then goes down continuously to the second, the stronger bond. Thus, if the hook rises to the peak of the loop, more of the force is applied to the weaker bond, which breaks. In this case, the stronger bond remains untested. It is apparent that the combination of a high-bond loop as well as a force distribution that tests the stronger bond is the reason why gold ball bonds are specified to have, and do yield, a higher pull force than aluminum wedge bonds. However, in both cases, a pull in the center of the loop for single-level bonding would provide a more reliable quality control for the total bonding process.

Gold ball bonds generally yield a higher pull strength than aluminum wedge bonds for the reasons cited. But ultrasonic gold ***wedge*** bonds made with wire having the equivalent breaking force and elongation of aluminum wire, yield equivalent pull forces to aluminum wedge bonds when the bond deformation is in the low-to-medium range (<2 wire diameters), as shown in Figure T-5. At higher deformations, however, aluminum ultrasonic bonds become metallurgically overworked, which weakens the heel region and lowers the pull force dramatically, often by a factor of 2, as the bond deformation increases above two wire diameters. The different metallurgical characteristics of gold wire permit deformations up to about 2.5 wire diameters with little decrease in the pull force. Unfortunately, dependence of the pull force on bond deformation is not recognized in some specifications. MIL STD 883, Method 2010, permits bond deformations of 3 wire diameters for both gold and aluminum wedge bonds.

1.2.3 Effect of Wire Elongation on Bond Pull Strength (Large-Diameter Aluminum and All Gold Wire)[T-5]

As is evident from equations (T-1) through (T-3), the bond pull force is strongly dependent on the ratio of the loop height to bond spacing. The loop height will increase during pulling if the wire elongates significantly. Small-diameter wire made for ultrasonic wedge bonding, either aluminum or gold, usually has an elongation of less than 2% and this has little effect on the pull force. However, small-diameter gold thermocompression bonding wire or aluminum wire bonds that have been annealed (e.g., during high-temperature glass-ceramic package sealing) can have elongations of 5 to 10%, and large-diameter annealed aluminum bonding wire, up to 30% (see Appendix T-1). During bond pulling, the h/d ratio may increase significantly, yielding a pull force higher than one might expect if only the breaking load of the wire and the initial bond geometry were considered. This will be even more significant if the initial value of h was low.

Figure T-2 showed that loop height is an important factor in determining the bond pull strength. Thus, it is apparent that significant wire elongation during bond pulling will change the loop height and affect the magnitude of the pull strength. Figure T-6 gives the loop height change versus elongation for three different bond-to-bond lengths, all starting with the same

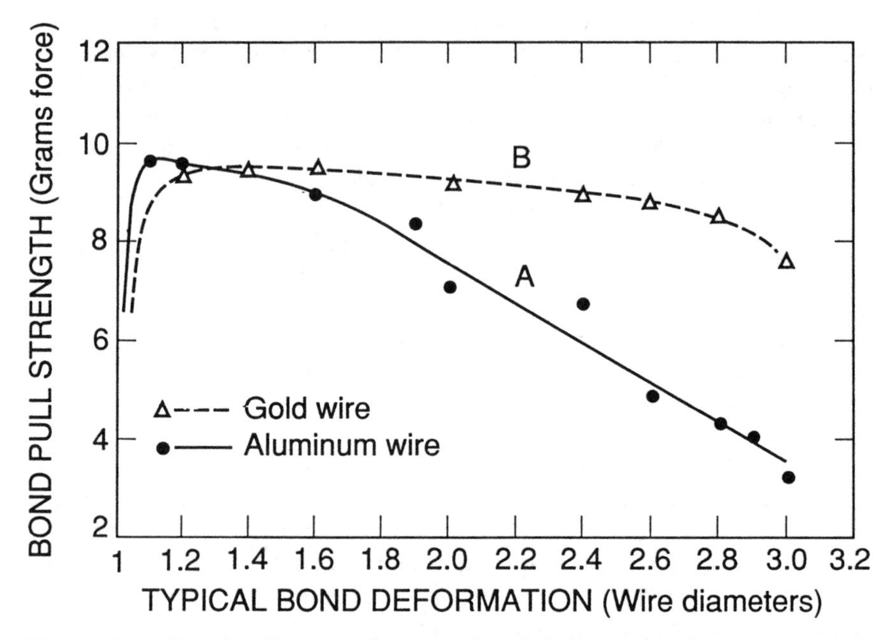

Figure T-5. Bond pull strength versus bond deformation for 25-μm (1-mil) diameter aluminum and gold wires, both having 13-gf breaking strengths. All bonds were made on the same bonding machine using the same bonding tool. Other bonding parameters were optimized for each metal to produce the best overall pull strength and the lowest standard deviation. All bonds were made on a single level. The loop heights were approximately 0.3 mm (12 mil) and the bond-to-bond spacing was 1 mm (40 mil). The loop-height-to-bond-spacing ratio is much larger than that generally found in device production. Thus, a scaling down of the bond pull strength axis by a factor of about two would be more typical of values obtained from integrated circuits (after Harman [T-5]).

initial loop height. The geometries were chosen to cover those often encountered in medium-to-high-power transistors, but they can be linearly scaled down to appropriate microelectronic dimensions as long as the ratio of loop height to bond spacing is kept constant.

Figure T-7 shows the effect of this wire elongation (incorporating the resulting loop height increase) on the bond pull strength, assuming the same initial geometry as in Figure T-6. In this calculation, all bonds break when the force in the wire reaches 500 grams. (See Appendix T-1 for actual properties of such wire.) For simplicity, the calculation was made for single-level bonds. From Figure T-7, it is apparent that the tendency of the bond pull strength to decrease with decreasing wire breaking strength

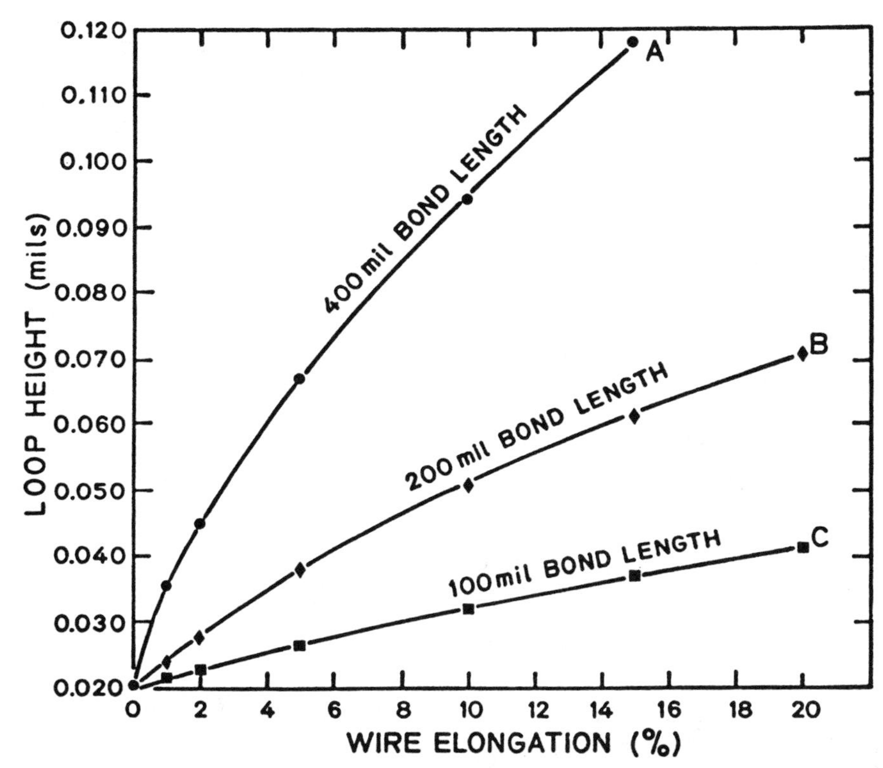

Figure T-6. The effect of wire elongation on the final loop height at the point of wire rupture during a bond pull test. All bonds had the same initial loop height, 0.5 mm (20 mil), but three bond-to-bond lengths were used as indicated.

can be partially offset by the increase in bond pull strength if the wire has increased elongation. Therefore, the pull strengths tend to be independent of the specific wire breaking strength for many common device geometries. These results may be scaled down for integrated circuits, except that small 25-μm (1-mil) diameter annealed wire does not elongate more than approximately 10%, and this only after high temperature exposure such as CERDIP sealing at >400 °C.

There will be cases where the effect of large wire elongations can significantly change the bond-pull geometry, and hence the measured pull strength, if the pulling probe (the hook and arm) is flexible or is free to pivot where it is joined with the force gauge or load cell. This may occur even if the hook does not slip in the case where a large post height is involved. Here, the wire span will be cònsiderable longer on the chip side

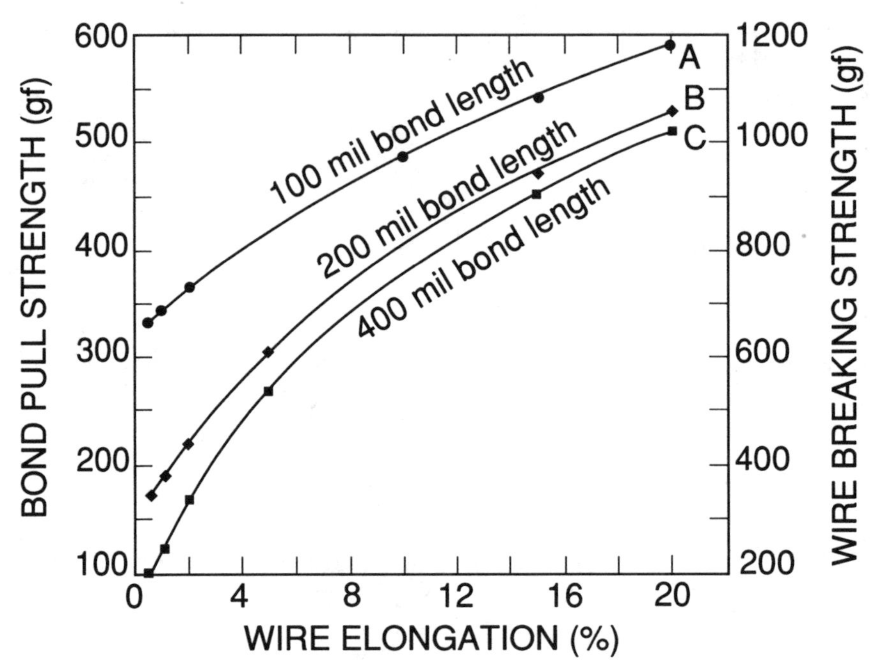

Figure T-7. Bond pull strength and wire breaking strength versus wire elongation. The curves are bond pull strengths based on the geometries of similar lettered curves in Figure T-6.

of the hook than on the post side. The relatively greater increase in length of the chip-side span during the test will result in moving (swinging) the hook nearer the post and in pulling on the wire at some angle Ø from the vertical. The effect will be enhanced if the pulling hook was initially placed nearer to the post than to the chip bond. These changes in the bond pull geometry, which can result in lower measured values of pull strength, must be taken into account in any pull test calculations involving wire with large elongation.

Stress-strain type measurements have been made during pull testing on a number of large-diameter power-device wire bonds in order to determine any unique characteristics that could influence the pull test. Both the measurement and its interpretation are much more difficult for pulling a typical wire-bond loop than for measuring the stress-strain relationship of a long piece of wire. In pulling a 250-mm (10-in) standard length of wire, the elongation is normally read directly from the recorder chart (see section 1.3, Figure T-9, for examples of stress-strain curves). However, in pulling a large-diameter wire bond loop, the total length of wire is generally less than 6.25 mm (0.25 in), and in addition the measurement indicated by the apparatus is in reality the increase in loop height (which is nonlinear with wire elongation) and is very small compared to the elongation of the standard length of wire. Thus, when determining wire-bond-loop elongation, the sensitivity of the measurement apparatus must be increased to its maximum, and any system nonlinearities, such as a slight irregularity of the screw thread pitch on the stress-strain machine or bending of the pulling hook, will have a greater effect and must be corrected for in each curve.

A typical, corrected force versus rise-in-pulling-hook curve for an 200-μm (8-mil) diameter emitter wire-bond from a power device is shown in Figure T-8. There are three distinct regions in this curve. Region (1) is the triangular loop formation and elastic wire-tensioning region. Although the curve increased linearly for this bond, other bonds often showed variations as the loop formed into a triangle, generally within the dotted curves. Point (2) denotes the elastic limit of the wire, and (3) is the region of inelastic (plastic) elongation, which in this case begins at approximately 60% of the bond pull strength. At point (X), the wire necks down rapidly and then breaks at (4). The elongation of the wire in region (3) was determined to be 10.5%, by using data from Figure T-4 and the measured bond geometry. More explanation, as well as stress-strain curves of typical bonding wires, is given in the next section.

1.3 NONDESTRUCTIVE PULL TEST [T-9]

1.3.1 Introduction

The nondestructive wire-bond pull (NDP) test is a variation of the destructive pull test in that the maximum applied force to the bond loop is limited to a predetermined value. It is usually applied on a 100% basis to all wires in a hybrid (or IC) or at least in areas found to have repeated bonding problems. The test is intended to reveal weak bonds while avoiding damage to acceptable bonds.

Evaluation studies of the NDP test method have been carried out by a number of organizations, but often the information was obtained for in-house purposes and remains unpublished. The several published reports [T-1, 6 to 8] indicate that the NDP test is valid under the specific conditions of each particular experiment. Of these, only Polcari and Bowe [T-1] recognized and discussed the importance of bond geometry. They also repeatedly stressed a number of bonds to their chosen NDP force and found that some did not fail during 100 applications of this force, whereas others failed after only four or five trials. The average bond withstood

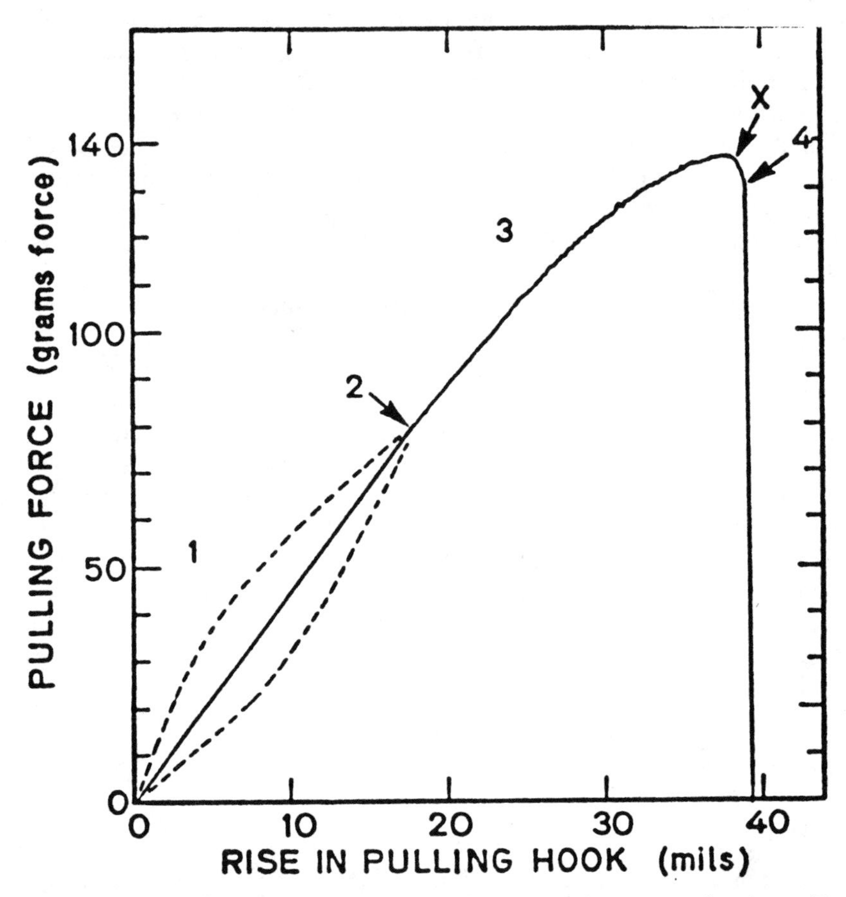

Figure T-8. Bond pulling force (related to stress) versus rise in pulling hook after wire contact (related to strain and elongation) for an 200-μm (8-mil) diameter aluminum wire bond on a power device. (1) is the triangular loop formation and elastic wire tensioning region; the dotted lines indicate the typical variations that are observed in this region; (2) is the wire elastic limit; (3) is the region of inelastic wire deformation; (X) is the region where the wire necks down rapidly and then breaks at (4). This curve was corrected for measuring apparatus nonlinearities.

about 50 successive applications of force before failure. However, the standard deviations of the destructive bond pull strengths of the bonds available to them were quite large. Many of the bonds would have been stressed beyond their elastic limits (see section 1.3.2). All of the bonds were stressed at forces higher than those recommended in the present work.

The nondestructive pull force is usually specified for a given wire diameter and metallurgy (see MIL STD 883, Method 2023). Typical values for 25-μm (1-mil) diameter wire are 2 gf for aluminum and 2.4 gf for gold. Various in-house specifications have ranged from 0.8 gf to 3 gf for the same wire size. However, specific values make no allowance for bonds having widely different geometries. The test can break a strongly welded wire bond if, because of package or other limitations, it must have a very low loop. Likewise, fixed pull values apply relatively less testing force to bonds with high loops.

A more scientific approach to deriving the NDP force is to consider the metallurgical characteristics of the wire in addition to the bond geometry. Figure T-9 shows two differing elongations of wire used for bonding. Although the curves are for aluminum wire, equivalent data for gold wire would be similar. Note that gold wire for use in thermosonic and thermocompression bonding is annealed and would generally have stress-strain characteristics nearer to those of curve A in Figure T-9. In order to avoid metallurgical change or damage to wires during pulling, the wire must not be stressed beyond its elastic limits, region 1 of the stress-strain curves.

1.3.2 Metallurgical and Statistical Interpretation of the NDP Test

The metallurgical and statistical interpretation of the NDP test was given by Harman [T-9]. A normal-distribution destructive bond-pull control limit of $\overline{X} - 3\sigma$, often used in the electronic industry, assures that only one normal-distribution bond out of 740 will have a pull strength below that force. Reducing the NDP force 10% to $0.9\,(\overline{X} - 3\sigma)$, where $\sigma \leq 0.25\,\overline{X}$, will assure that no bond within the normal $(\overline{X} - 3\sigma)$ distribution is stressed past its elastic limit, whereas any freaks (bonds with low, non-normal, bimodal, etc., pull strengths) will be weeded out. Only those bonds whose pull strength lies in the range of $(\overline{X} - 3\sigma)$ to $0.9\,(\overline{X} - 3\sigma)$ may be stressed, to some degree, beyond their elastic limits.

All bonds with pull strengths below that range will be broken, and all bonds with pull strengths above it will only be stressed within their elastic limits. The actual percentage of bonds that lie within the inelastic stress range will depend on the relationship between $\overline{X}$ and σ. As an example, if $\sigma = 0.25\,\overline{X}$, only one normal-distribution bond in approximately 2600 lies within that range, and if $\sigma = 0.15\,\overline{X}$, only one bond in approximately 1000 lies within that range. Most bonds made on average production lines have standard deviations within these limits.

In cases where very low standard deviations are encountered ($\sigma \leq 0.15\,\overline{X}$), as may happen with automatic bonding machines on integrated circuits, the

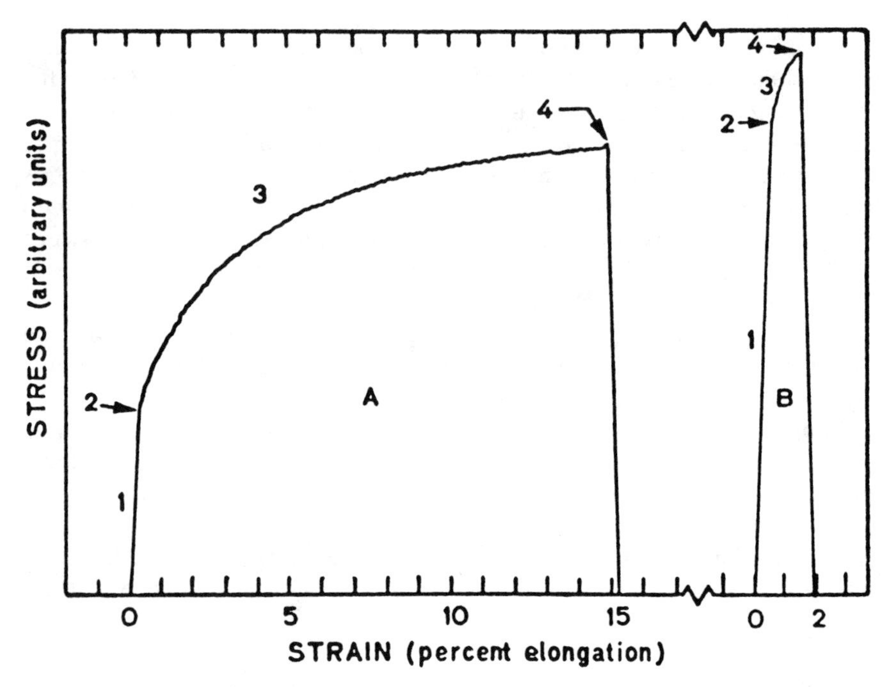

Figure T-9. Typical stress versus strain (elongation) curves for 99.99% pure aluminum bonding wire in two states of hardness: wire (A) is annealed and is typical of large-diameter aluminum bonding wire; (B) is stress relieved (partially annealed). Its characteristics are similar to that used for small diameter ultrasonic wedge bonding. In order to display both curves on the same chart, the stress axis was made arbitrary (the breaking load of (A) was approximately one-half that of (B)). On both curves (1) is the elastic region where the stress is proportional to the strain, (2) is the proportional or elastic limit, (3) is the region of inelastic or plastic deformation, and (4) is the breaking load of the wire. The elongation at the breaking point is 15% for (A) and 1.5% for (B). Some modern Al, 1% Si wires have a flatter region (3) than shown.

NDP force can be changed to $0.9\ (\overline{X} - 4\sigma)$. In this case, no more than one normal-distribution bond out of approximately 30,000 would be stressed past its elastic limits. In a situation where $\overline{X} = 6$ gf and $\sigma = 0.15\ \overline{X}$, the NDP force would be approximately 2.1 gf, and only one normal-distribution bond in 45,000 would be stressed beyond its elastic limits. Table T-1 gives the percentage of those bonds whose pull strength will lie in the inelastic stress range for both the normal and the low σ criteria.

For low elongation wire, the maximum safe NDP force is $0.9\ (\overline{X} - 3\sigma)$, where $0.25\ \overline{X} \geq \sigma > 0.15\ \overline{X}$, and $0.9\ (\overline{X} - 4\sigma)$ when $\sigma \leq 0.15\ \overline{X}$. No NDP

TABLE T-1
PERCENTAGE OF BONDS IN THE INELASTIC STRESS RANGE*

Standard Deviation as Percentage of $\bar{x}$	Percentage of Bonds with Pull Strengths Lying in the Range	
	$(\bar{x} - 3\sigma)$ to $0.9(\bar{x} - 3\sigma)$	$(\bar{x} - 4\sigma)$ to $0.9(\bar{x} - 4\sigma)$
25	0.038	
20	0.066	
15	0.1	2.2×10^{-3}
10	0.12	3.0×10^{-3}
5	0.13	3.2×10^{-3}

* This table is calculated on the assumption that the bond pull strengths (excluding freaks, which usually have very low pull-strengths) fall approximately within a normal distribution rather than for example a bimodal one. If more bonds than predicted have pull strengths falling below that of a normal distribution, particularly in the range of $\bar{x} - 3\sigma$ to $0.9(\bar{x} - 3\sigma)$, then more bonds may be damaged than are indicated in the table, and conversely. Plots of bond data on normal probability paper can be used as a simple means of determining the normality of the distribution.

testing is recommended for cases where $\sigma > 0.25\ \overline{X}$ since this indicates that some aspects of the bonding procedure are out of control and either a low, meaningless NDP force would have to be used or too many bonds would be stressed beyond their elastic limits and/or broken. Table T-2 gives a summary of the NDP test recommendations for wire with various elongations.

1.3.3 Assessment of Any NDP Test-Induced Metallurgical Defects

During the NDP test, using the NDP force limits derived above, the wire is only subject to approximately one metallurgical stress-fatigue cycle. Bulk aluminum and gold will normally withstand hundreds of thousands of such cycles when the stress is kept below the elastic limit. The stress during the NDP test is primarily along the wire; thus there are essentially no outer-fiber-strains (from bending) in the bond heel area to enhance the probability of nonannealable crack formation.

Under these conditions, any stress-fatigue developed below the elastic limits of the bond-loop system during the NDP test should be small. Also, almost all devices whose reliability is critical enough to require NDP tests will routinely undergo thermal screens, such as burn in ($\approx$125 °C for 168 hours or equivalent) or such screens could be added if desired. These screens should be adequate to anneal any threshold level of NDP-test-induced fatigue occurring below the elastic limits, and they can also anneal some, if not all, of the stress-fatigue which might occur above the elastic limit, assuming no crack has formed [T-10]. Thus, only a small fraction of the NDP-tested bonds, whose strength is in the inelastic stress range of Table T-1, would retain a significant number of test-induced metallurgical defects after a typical burn in or other annealing period. Even for a case where a small nonannealable crack remains in the bond heel, it would be detrimental to the subsequent operating life of the device only if severe vibrations (such as ultrasonic cleaning) were encountered or if the device was subject to temperature cycling and then only if the bond loop was small (see section 5.4).

Twenty years after the invention of the nondestructive pull test at Autonetics (Rockwell), the idea of the nondestructive pull is still controversial. Some people are worried about metallurgical damage to the neck or heel of the bond, and others are concerned that the hook might hit and damage an adjacent wire as it is being positioned.

At this writing (1989), there have been at least a hundred million nondestructive wire pull tests [T-11], which are a requirement for all military class-S devices. All of the evidence available indicates that the test is truly nondestructive. In addition, Blazek [T-12] has shown that the NDP test

TABLE T-2
SUMMARY OF NDP FORCE RECOMMENDATIONS

Type of Production	Wire		Relation Between $\overline{x}$ and σ on the Bond Pull Test	NDP Force Recommendation
	Composition	Elongation		
Normal	Aluminum	< 3%	$(0.25 \leq \sigma > 0.15)\overline{x}$	$0.9(\overline{x} - 3\sigma)$
High rel.	Aluminum	< 3%	$\sigma \leq 0.15\overline{x}$	$0.9(\overline{x} - 4\sigma)$
All	Aluminum	~ 5 to 20%	$\sigma \leq 0.25\overline{x}$	$\frac{\overline{x} - 3\sigma}{2}$
	Aluminum	> 20%	$\sigma \leq 0.25\overline{x}$	$\frac{\overline{x} - 3\sigma}{3}$
All	Gold	Use same elongation and σ rules as aluminum, except that the elastic limit is less predictable from one manufacturer to the next.		

does not lower the bond strength distribution of devices that later undergo the usual military qualification tests of temperature cycle, burn in, shock, and vibration. With regard to damage to adjacent wire bonds, a trained operator is less apt to damage a wire with the hook while positioning for a pull than an equivalently trained operator is to misplace or otherwise damage a wire while actually making a bond. Automatic nondestructive pull testers are currently available that are designed specifically to avoid touching adjacent wires.

The nature of immature and otherwise poorly bonded interfaces has been described by Harman [T-13]; see also Appendix IA-1. They consist of a series of unconnected microwelds. When an appropriate force is applied, the interface begins to separate, resulting in a "crack." This "crack" propagates rapidly with characteristics similar to those of a "Griffith crack" and completely breaks within a few milliseconds [T-13]. If the force is below a threshold value, then no break or damage occurs to the interface. Thus, the NDP test is largely a go, no-go test, and any possible marginal damage can be assessed by the statistical methods outlined above.

1.3.4 Limitations of the NDP Test

Regardless of all comments above, the user of NDP-tested devices must be aware of the limitations of this test. The test will only perform one function. It will remove weak, poorly made bonds with pull-strengths below the chosen pull-force level at the time the test is ***performed***. It is no assurance against later bond strength degradation due to gold-aluminum intermetallic and subsequent void formation, halogen corrosion, ultrasonic-cleaner-induced wire-bond vibration fatigue, or wire bond flexure fatigue due to temperature or power cycling, etc. (for a discussion of such possible failure mechanisms, see section 5.4 on mechanical failures).

The effect of post-NDP-test screens and environments on bonds should be thoroughly understood by the device user. These are discussed elsewhere in this book. In some cases, the devices can be chosen or designed to minimize post-NDP-test degradation, such as using monometallic wire and bond-pad systems and using high bond loops. The NDP test, as well as the destructive pull test, is not appropriate for screening the quality of ball bonds (see section 1.4, Ball Bond Shear Test). The welded area of a ball bond would have to be less than approximately 10% of the interfacial area to fail during a standard NDP test (pulled at 3 gf for a 25-μm (1-mil) diameter wire). This could only happen if the bonding process was completely out of control.

1.4 BALL BOND SHEAR TEST [T-14]

1.4.1 Introduction

The wire-bond pull test is universally used to assess the strength and to determine bonding machine setup parameters of wire bonds used in microelectronics (see section 1.2). Often, technicians and engineers assume that pull test data, which are adequate to determine wedge bonding machine setup parameters, are also sufficient for setting up the ball-bonding machine parameters. However, considering that most ball bonds can have interfacial welded areas in the order of 6 to 10 times the cross-sectional area of the wire, it is apparent that the wire will break in pull testing before even a poorly welded ball will lift.[3] In addition, the wire near the neck of the ball bond is totally annealed, recrystallized, and becomes the weakest part of the ball-bond wire wedge-bond system. In a pull test, the wire often breaks at this point, depending on where the hook is placed and the bond geometry (see section 1.2). Thus, little information is gained on the strength of the ball-to-bonding-pad interface if the ball is welded over more than 10% of its interfacial area. Considering the above, it is apparent that some type of ball-shear test offers the best possibility of properly setting up a ball bonding machine or of assessing the quality of a ball bond.

The ball-shear test was independently introduced to the microelectronics industry in 1967 by Arleth and Demenus [T-15], and Gill and Workman [T-16]. However, it appears to have been ignored or forgotten for almost 10 years until Jellison [T-17,18], and later, Shimada et al [T-19] designed precision shear testers and used them in a series of laboratory experiments that clearly demonstrated the usefulness of the test. Since that time, there have been numerous studies of the ball-shear test as well as applications of it to production quality control. In the past, there were two impediments to widespread implementation. One was the lack of commercially available, ***easy-to-use, and dedicated*** ball-shear equipment; the other was the lack of standards and military requirements for its use. At the present time, both of these impediments have been removed.

[3]Consider a ball bond on a 25-μm (1-mil) diameter gold wire. The diameter of a well-bonded ball is usually in the 65- to 90-μm (2.5- to 3.5-mil) diameter range, and the wire-to-bonded-area ratio is from about 6 to 10 in favor of the ball. Thus, in a wire-bond pull test, the wire will break even if the ball is weakly welded. Arguments similar to this were first pointed out by Gill and Workman [T-16], and variations of them have appeared in most of the later studies involving the ball-shear test. For instance, Stafford [T-20] calculated the force that the wire can apply on the ball-metallization interface in the pull test and has shown the above conclusion quantitatively.

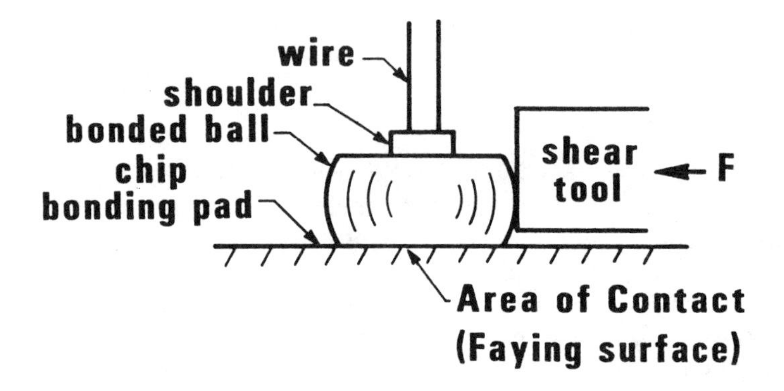

Figure T-10. Schematic drawing of the ball shear test. The bonded (welded) area may be less than the faying area (area of intimate contact). The typical outside diameter of a bonded ball from a 25-μm (1-mil) diameter wire is from 75 to 110 μm (3 to 4.5 mil). The height of the ball above the bonding pad is usually less than 25 μm (1 mil).

1.4.2 Apparatus

Quantitative Instruments

The equipment used to perform the ball-bond-shear test has ranged from tweezers and other hand-held probes to dedicated shear-test machines with strain-gauge force sensors and various electronic methods of data presentation. In principle, the ball-shear test is simple and consists of bringing some form of a shear tool up to the side of a bonded ball, applying a force sufficient to push it off, and recording that force. The test is illustrated in Figure T-10.

Jellison [T-17] and Shimada et al [T-19] designed similar precision mechanical systems with strain-gauge force readouts. The details of Jellison's design are appropriate for any dedicated ball-shear tester. A sketch of his apparatus is shown in Figure T-11. His tester employed a rigid, low-friction, linear bearing to transmit the load from the tool to the strain gauge. The sample was placed in the horizontal position and viewed from above with a microscope. The shear tool extended downward from its clamp so that it could operate in a deep package. A motor drive moved the test sample work holder to perform the shear test at a fixed rate of 0.2 mm/s (8 mil/s). This rate is not critical since Panousis and Fischer [T-21] have shown the rate of shear force application does not affect the shear force value over the range of from 0.13 to 3.3 mm/s (5 to 130 mil/s).

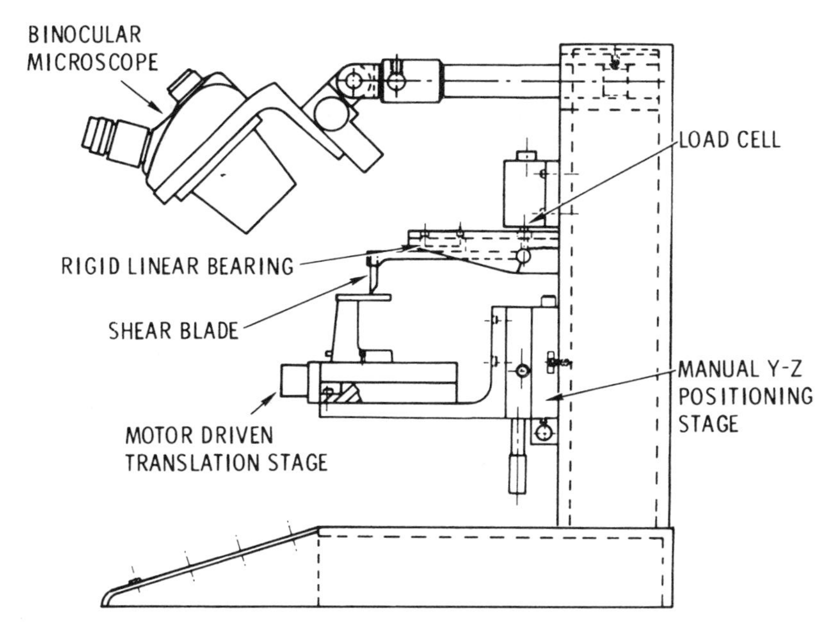

Figure T-11. Sketch of a precision microshear tester (after Jellison [T-17]).

Currently, there are several commercial ball shear testers available. The most sophisticated of these[4] have automatic vertical position finders, record data and failure modes, and can be interfaced with a computer for final data analysis.

1.4.3 A Manual Shear Probe as an Aid in Setting Up a Ball Bonder

It is desirable to have a quantitative, precision shear tester. However, if such is not available, a simple substitute can be made that will give information which, while not quantitative, will nevertheless allow one to quickly set up ball bonding machine parameters and qualitatively evaluate bonds for other purposes.

The simplest and most readily available tool for manually pushing off ball bonds is the blunted end of one tine of a tweezer and many production personnel have used them for this purpose for years. However, tweezers are relatively awkward to use, particularly if the ball is strongly attached. A simple manual shear probe was designed by Harman [T-14]. Figure T-

[4]Dage Precision Industries, Inc., 2372 Walsh Ave., Santa Clara, CA 95051. J. & A. Keller Machine Co., Inc. 2320 Military Rd., Tonawanda, NY 14150.

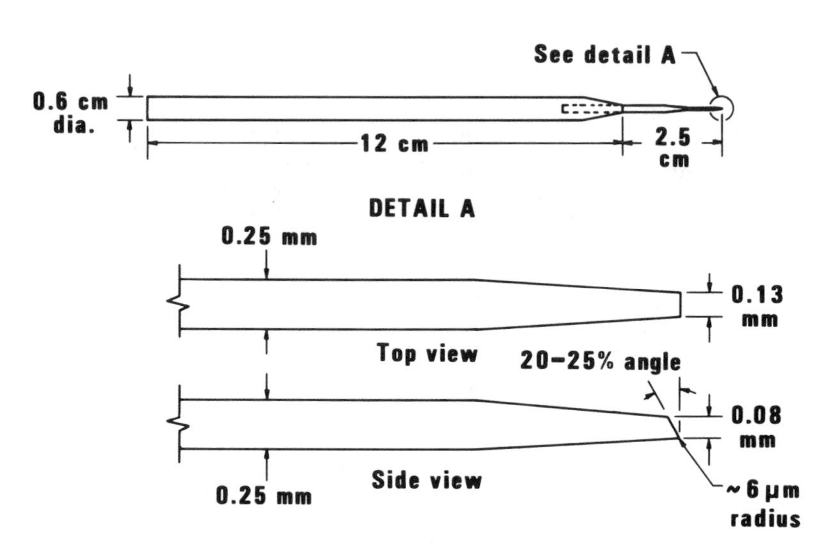

Figure T-12a. Detailed sketch of manual ball shear probe.

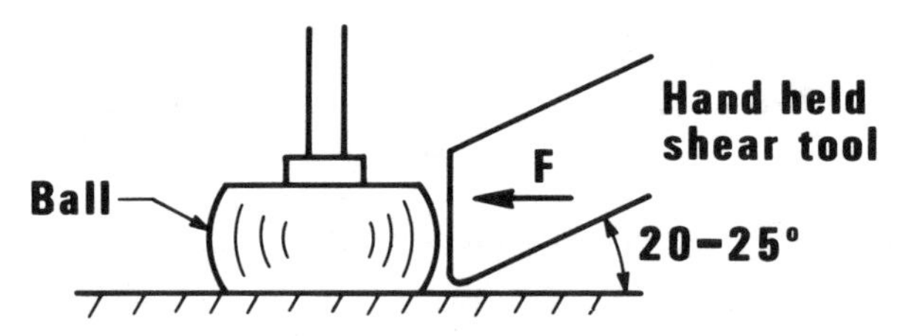

Figure T-12b. Sketch of manual shear probe in use.

12a is a drawing of a probe tip that is appropriate for use on balls made from wire of up to 33 μm (approximately 1.3 mil) in diameter. The tip can be made by a machinist or a precision tooling company. Currently (1989), a similar shear probe is offered by at least one bonding-tool company. However, a reasonably satisfactory one can be made in a few minutes using the smallest blade from a standard jeweler's screwdriver set. This blade, while still in the jeweler's screwdriver as a holder, can be manually narrowed and thinned on very fine emery paper to the approximate dimensions of Figure T-12a.

For use, the sample is placed on a holder at a height such that the probe, when held like a pencil, will approach the surface at an angle of about 20 to 25 deg. This matches the angle on the tip of the tool. The ball bond is then contacted perpendicular to its radius as shown in Figure T-12b. It is often useful to stabilize the hand holding the probe with the other hand. A binocular microscope with no less than 30X magnification should be used. Practice should start with strong ball bonds. These will generally have a shear force greater than 50 gf. For comparison, some weak balls can be made by using the same bonding machine settings as for the strong ones but locating half or more of the ball off the bonding pad. If the pads are on an IC, this will place part of the bond on glass where it will not weld, reducing the shear strength proportionately.

An indication of the bond strength can often be obtained by observing the deformation or smearing of the ball. (See Figure T-13 in the following section as an example.) Strong gold balls on gold metallization usually smear. However, strongly welded balls on aluminum sometimes come off suddenly at the peak force and are relatively undamaged. Also, balls that are relatively high 38 μm (approximately 1.5 mil) are often left relatively undamaged during the test. Bond evaluation using a manual probe usually takes place by dividing bond quality into three nominal strength groups. For average size balls on 25-μm (1-mil) diameter gold wire, the shear force of the poorest group is in the range of <20 gf, intermediate, 25 to approximately 40 gf, and strong, over 50 gf. (The values will be lower if small-diameter balls are tested; see Figure T-15 for expected values.) These force values are separated enough for a trained operator to "feel" the difference. An instrumented manual probe showed that reasonable comparisons can be obtained manually [T-14], so if one were commercially available, it would be useful for quantitatively setting up machines.

1.4.4 Interferences to Making Accurate Ball Shear Test Measurements

As with any test method, there may be problems in performing the ball shear test that can produce incorrect or misleading data.

Shear Tool Drag. One of the most common problems that occur is the improper vertical positioning of the tool. The tool should not drag on the substrate. It should approach normally deformed balls from 2.5 to 5 μm (approximately 0.1 to 0.2 mil) above the substrate and for large, high balls, no higher than 13 μm (approximately 0.5 mil). (The bottom of the tool must be kept clean to permit such positioning.) If the tool is positioned higher, it will tend to ride-over or smear-over the top of the ball, depending on the height of the ball. If substrate dragging occurs on thin films, then the indicated shear force can increase by 10 to 20 gf. Some chips are not attached horizontally, and additional care must be taken to prevent the shear tool from contacting the bond pad metallization during the test.

Gold-to-Gold Friction Rewelding, Shearing gold bonds on gold substrates can lead to an unusual interference. Gold is capable of friction welding to gold surfaces. A SEM photograph of a deformed gold ball and gold bonding pad showing the results of multiple friction rewelding is given in

Figure T-13a. An x-y recorder plot of shear force versus time for an initially strong ball bond on a 33-μm (1.3-mil) diameter gold wire that underwent similar multiple friction rewelding is given in Figure T-13b. The first (highest) peak was the initial ball shear, and the two following peaks occurred during successive friction rewelds. The dotted line indicates the normal shape of the force versus time curve when rewelding does not occur.

Rewelding problems may be eliminated by spraying a very thin layer of solvent-thinned oil on experimental substrates, after bonding but before shear testing. Charles [T-22] designed a tool that was slightly ground back from the forward edge, which lifted the ball, and prevented rewelding. He also noted that if data are taken with a peak-reading system, the initial shearing of well-welded balls produces the highest force value, and subsequent rewelds are lower and thus not recorded. However, there is still a possibility that intermediate strength bonds may be equaled or exceeded in shear strength by rewelds. Friction rewelding seldom occurs while shearing bonds on aluminum metallized IC pads, where the small size of the pad, as well as the aluminum oxide surface film presumably prevents it.

Interferences When Shearing Bonds on Thick Films. There are several potential interferences that can occur when shearing ball bonds on thick film metallization. One would normally assume that balls of a given diameter would yield shear forces somewhat lower when bonded to thick films than when bonded to thin films, since thick films contain pits and voids and in some cases glass occlusions on the surface. There have been several shear test studies of ball bonds on thick film metallization in the literature [T-23,24]. However, there was not enough information given on the welded ball size, or the actual welded area, to compare directly with the extensive data published on thin films.

One might expect to use the same experimental procedure in shearing bonds made to thick films as for thin films. However, the thick films themselves are often higher than the vertical position of the tool above the substrate when set for shearing bonds on thin films. Therefore, tests made on thick film hybrids may present serious problems in vertical positioning of the tool. It is therefore possible that the tool will be improperly positioned and drag across the thick film. This will result in a very high apparent value for the ball-shear force and may explain why some reported values of ball-shear force from thick film tests are higher than the shear strength of gold [T-23].

Even if the vertical positioning of the shear tool is correct, shear tests made on bonds welded to thick films, as well as to thin films, can yield much lower shear values than expected if the metallization adheres poorly to the substrate [T-16]. An example of shearing a ball on a poorly adherent

Figure T-13a. An example of a gold ball strongly bonded to gold metallization that underwent friction rewelding to the pad. Note also the gross deformation of the ball resulting from the shear test. This TS bond was made from 25-μm (1-mil) diameter wire (after Weiner et al [T-25]).

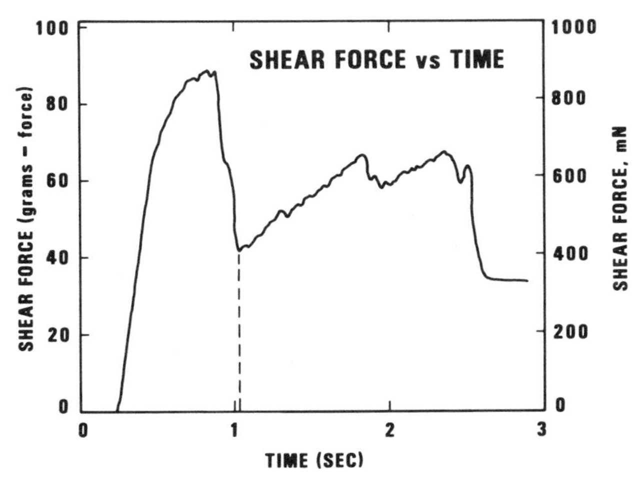

Figure T-13b. A x-y recorder plot of shear force versus time for strongly welded 32-μm (1.3-mil) diameter gold wire ball bond on a clean plated gold substrate. The dotted line indicates the final break of the ball that does not reweld during the shear test. Successive peaks to the right of the dotted line are the result of friction rewelding.

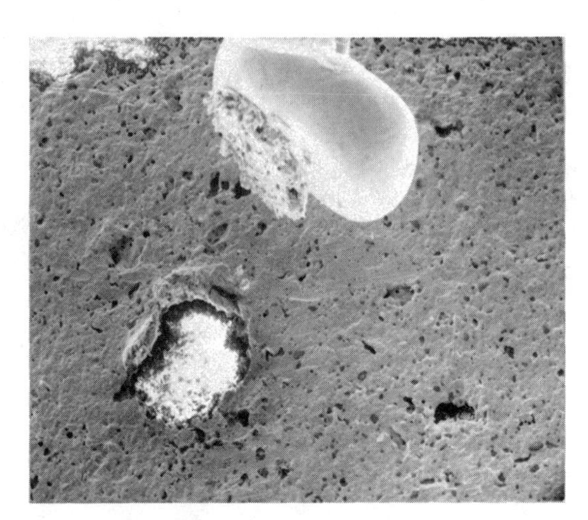

Figure T-14. An example of a TC ball bond sheared from a poorly adherent gold thick film. This ball sheared at 24 gf, whereas equivalent sized bonds to *adherent* thin-film gold sheared at about 80 gf. The bond was made from 32-μm (1.3-mil) diameter gold wire.

thick film is given in Figure T-14. In addition, when bonding to pads on semiconductor chips, it is possible to fracture (crater) or at least damage the semiconductor. Application of the shear test to balls that damaged and cracked the silicon during bonding may result in cratering and low shear forces during the test. ***The shear test can therefore be used to evaluate bonding machine setup parameters, to minimize semiconductor damage (cratering) not detectable with the pull test, as well as to test metallization adherence.*** See the section on cratering for a more complete discussion of this problem.

Interferences When Shearing Compound Bonds (Ball on Ball). There has been only one published paper on a shear-test evaluation of compound-thermosonic bonds. Schultz and Chan [T-39] found that centering of the top ball bond on the bottom ball was very important. They found an increased tendency to crater (as revealed by the ball-shear test) in cases where the top ball bond was misaligned on the bottom one. Some bonding machines resulted in more craters than others without apparent reason. Increased ultrasonic energy also significantly increased cratering. However, Schultz and Chan found that by careful machine adjustment, the energy could be minimized and essentially cure the cratering problem. When optimized, and centered, the bond shear strength of the lower ball was statistically unchanged from that of a single ball bond.

One must be concerned with the increased cratering probability, especially on devices that are known to be crater-prone. Therefore, unless more definitive studies reveal the unknowns in the process, the use of compound bonds (which is permitted by MIL STD 883) should be minimized.

Cleanliness of Shear Tools. Since the shear tool should be positioned within less than 13 μm (0.5 mil) from the chip or substrate, its bottom surface must be kept clean. Metallic smears, bits of aluminum, gold, glassivation, silicon, etc. can stick to the bottom and prevent proper vertical positioning. This problem is most significant when shearing very flat (low profile) balls with tools that are several times wider than the diameter of those balls. A narrow tool (about 1 ball diameter wide) tends to clean itself off as it shears through metal.

1.4.5 Ball-Shear Strength Versus Bonded Area

Experimental shear strength data versus bonded area for microelectronic ball bonds have been published by two independent sources [T-18,19] and verified by Harman [T-14], who measured the shear strength of gold wire 3,630 kg force/cm^2 (13,100 psi). In addition to gold, various aluminum alloys with different strengths are used for bonding pads (e.g., pure Al; Al, 1% Si; Al, 3% Cu; etc.). These would have been subjected to various sintering (heat treatment) conditions. Because of these variations, shear values obtained from aluminum wires (whose diameter is comparable to the dimensions of a bond pad) can only serve as a general guideline for the shear strength of thin-film aluminum metallization. The ultimate shear strength for the hard and annealed 1% silicon, aluminum wires were 20,200 and 560 and 340 kg force/cm^2 (12,200 psi), respectively. The shear strength values of the hard and annealed aluminum wires bracketed those of gold, indicating that a ball bond weld can fail either on the aluminum or gold side depending on the particular characteristics of the aluminum pad. These data are all plotted in Figure T-15 as shear force versus diameter of the bonded area.

The wires used in this experiment contained Cu-Ag impurities. Gold wire used in automatic bonding machines is usually stabilized with beryllium which results in a 10 to 15% stronger gold. Such gold wire might give somewhat higher gold ball shear values or lead to more failures in the aluminum than in the gold side of the weld. Values obtained from the curve should be considered maximum-theoretically-obtainable values rather than expected values, since all of the area under a ball is not welded. Stafford [T-20] has estimated that a thermosonic ball bond to an aluminum film is only welded over about 65% of the area of a typical 75-μm (3-mil) diameter ball deformed 50% during bonding. This model has not been verified; nevertheless, the entire area under a bond is not expected to be

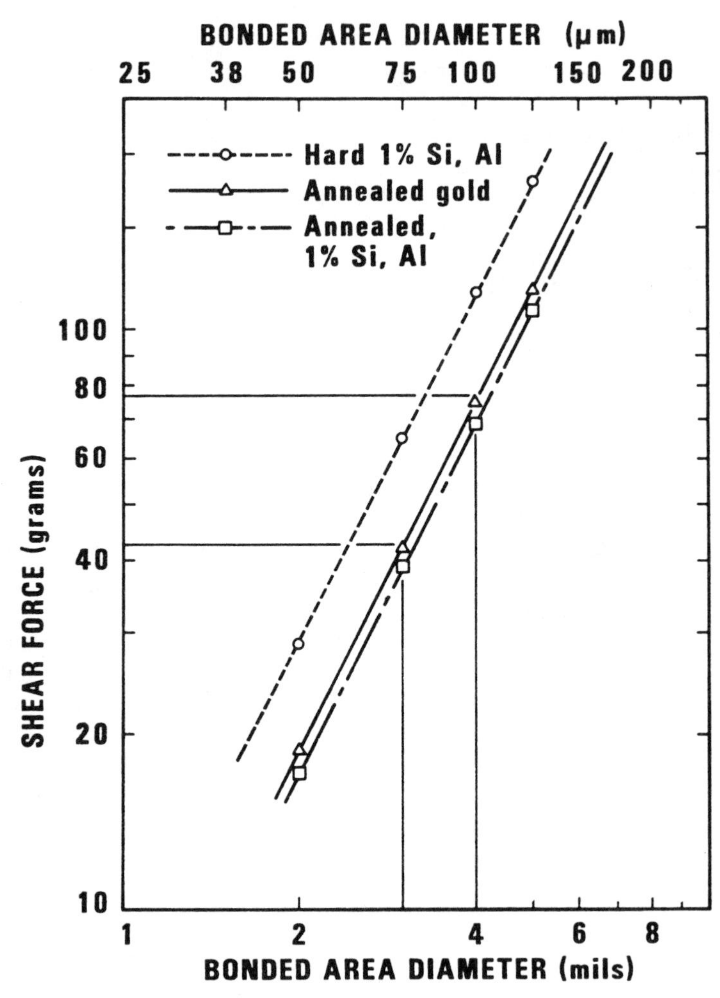

Figure T-15. Shear force versus bonded area for evaluating the maximum expected values to be obtained from the ball shear test. Only one curve is given for gold because the ball is annealed. However, since the joint will fail in its weakest member, the shear strength range of aluminum is given for guidance as to the possible strengths of aluminum metallization. For use, the diameter of the tool impression in the ball is measured with a microscope.

welded. In addition, one may expect different amounts of welded area for optimized bonds to clean metallization depending on the type of that metallization and the method of bonding. For instance, testing 75- to 90-μm (3- to 3.5-mil) diameter balls, Weiner et al [T-25] obtained shear force values of well-made bonds to gold metallization of approximately 40 gf (very near the value in Figure T-15) and to aluminum of approximately 30 gf (about 70% of values in Figure T-15 and close to Stafford's prediction). However, ball-shear values from gold TS bonds on aluminum supplied to this laboratory (NIST) by N. T. Panousis (Signetics) for an intercomparison, produced shear force values indicating that they were welded over essentially the entire ball diameter. Thus, the amount of welded area to be expected from optimum TS bonds to aluminum cannot be predicted at the present time and at specific production control values.

Observation of numerous cross sections of ball bonds at this laboratory (NIST) indicates that the outer boundary of the tool imprint on top of the ball corresponds closely to the perimeter of the bonded area. It is, therefore, recommended that the welded area should be estimated by measuring the outer diameter of the tool imprint in the ball rather than the outside diameter of the ball. This value should then be used to obtain the maximum expected shear force in Figure T-15.

The curves of Figure T-15 can be used to establish the maximum shear force obtainable in as-made ball bonds. However, no minimum acceptable value can be deduced from that curve. Panousis [T-21] has addressed this problem by comparing wire pull and ball shear data. He found that if the mean pull strength is plotted against the mean shear strength, a minimum shear force will be found in which no ball bond lifts occur during pull testing. Such a procedure requires thousands of bonds to obtain meaningful data. In his case, the mean shear force of good bonds was approximately 80 gf and the crossover shear force, where some ball-lifts occurred during pull testing, was 40 gf. Therefore, by this criterion, a minimum acceptable shear force appears to be approximately one-half the indicated value in Figure T-15.

1.4.6 Effect of Gold-Aluminum Intermetallic on Shear Strength

When gold is bonded to aluminum at elevated temperatures, intermetallic compounds will form. There may be a question concerning their effect on the shear test. There appears to be no published data on either the tensile or the shear strength of such compounds. However, Philofsky [T-26] made estimates of the tensile strength as a result of tensile testing Kirkendall void-free Au-Al couples and concluded that all of the intermetallic compounds are at least three times as strong as annealed gold or aluminum. Considerations of the binding energy of the compounds would suggest that these compounds could be 10 times as strong as either gold or aluminum, and this is generally verified by hardness measurements published by Kashiwabara and Hattori [T-28]. Even though such com-

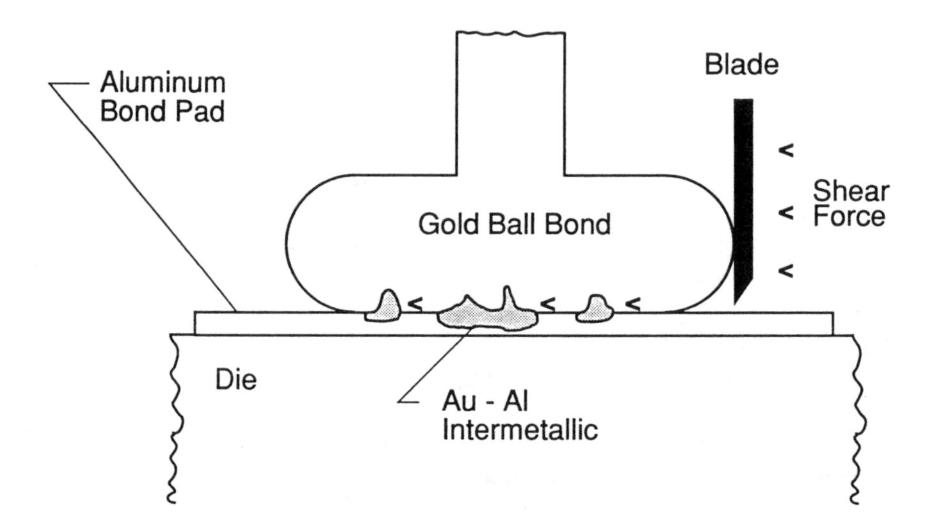

Figure T-16. Schematic of a ball bond with isolated intermetallic growths. These growths can offer considerable resistance to a ball-shear probe but can often be "pried" up with a scalpel or will lift in a pull test (after Devaney [T-27]).

pounds are brittle, we conclude that they should not result in lowering the ball bond shear strength as long as the interface is void-free. However, there is the possibility that void-free compounds could increase the strength since intermetallics usually form irregularly in depth across the interface and effectively increase the surface area of both the gold and the aluminum. Shimada et al [T-19] obtained such an increase in strength (about 10%) during the early part of his high-temperature tests, and this might explain why bonds made at relatively high temperature are reported to be the strongest, as discussed below.

1.4.7 Pluck Test

It was pointed out by Devaney [T-27] that if a gold ball bond is poorly made and subsequently undergoes thermal stress, intermetallic spikes will form which will extend into the gold. These spikes will add lateral strength to the bond when it undergoes shear testing and yield a deceptively high shear force. For failure analysis purposes, Devaney used a fine scalpel blade to pry or "flip" the bond up, leaving the intermetallic spikes on the pad for examination. He called this the "pluck test." Figure T-16 shows such a bond. Weak "as made" bonds would have been revealed by a shear test during production, and either the problems of poor bonding machine setup or contamination could have been resolved at that time. Thermal stress experiments on intentionally weak bonds (shear force <50% of optimum) made on clean pads by Harman (unpublished) indicate that when that "as-made" shear force has decreased to about half the original value, the Devaney mechanism becomes significant. In some cases, bonds with shear strengths of 10 to 15 gf will lift in a pull test at 3 to 5 gf. This happened frequently enough to require application of the nondestructive pull test at various stages of the ball-shear, thermal-stress experiments to remove bonds with this failure mechanism. It should be noted that strongly welded gold ball bonds to clean aluminum pads that are thermally stressed result in uniform intermetallic formation and have ***not*** been observed to fail by the Devaney mechanism.

Recently, Clatterbaugh [T-29] observed that the formation of intermetallic compounds under a ball bond can produce considerable stress on the silicon. The added stress of a ball shear test can then result in silicon damage (cratering). See the section on cratering for a discussion of this.

1.4.8 Comparison of the Ball Shear and the Bond Pull Tests

White [T-30] has published the most effective comparison between the ball shear and the bond pull test. Strong, gold ball bonds were made to aluminum integrated circuit metallization, and they were put on temperature test at 200 °C for 2688 hours. The degradation of the gold-aluminum interface was studied by monitoring both the shear and pull test at various time intervals. His data are replotted in Figure T-17. The bond interface strength decreased by a factor of 2.6, presumably due to Au-Al intermetallic formation and some Kirkendall voiding. However, the voiding was not sufficient to impair the electrical operation of the device, and there was no evidence of the Devaney mechanism [T-27]. During this time, the pull test actually indicated that the strength of the bond increased slightly, presumably due to changes in the wire metallurgy.

1.4.9 Applications of the Ball Shear Test

Bonding Machine Setup Parameters, Thermocompression Bonding. Directly or indirectly, the majority of published work on the ball-shear test has resulted in improving the set-up parameters of the bonding machine. Even investigators studying contamination and/or reliability found it essential to properly set up the bonding machine using the shear test before beginning their studies. Thus, Jellison [T-17] first studied the effect of various bonding forces and interface temperatures on the bond shear strength of gold TC bonds made on clean gold metallization in order to compare shear

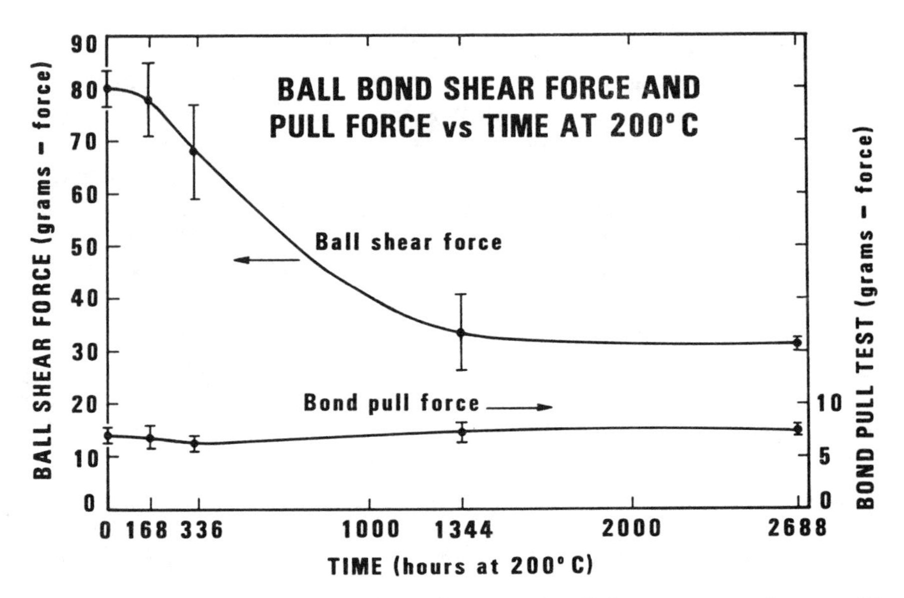

Figure T-17. Gold ball bond shear force and pull force versus time at 200 °C. The ball bonds were made with 25-μm (1-mil) diameter gold wire, bonded to integrated circuit aluminum bonding pads. Note the change of scale from shear force (left) to pull force (right). (Curve is a replot of data from White et al [T-30]. Actual data for this figure were supplied by A.W. Schelling.)

strength results with those results from contaminated surfaces. Figure T-18 is an example of this use of the shear test. Shimada et al [T-19] made similar observations for gold TC bonds on aluminum metallization even though they were primarily studying bond degradation effects of gold-aluminum (Au-Al) intermetallic compound. Combining their various data, one would choose a bond interface temperature of 300 °C, a bonding time of 0.2 s, and a bonding force of 100 to 125 gf to obtain strong thermo-compression ball bonds from 25-μm (1-mil) diameter gold-wire on either aluminum or gold metallizations. These parameters offer good shear strength even in the presence of a moderate amount of organic contamination.

Bonding Machine Setup Parameters, Thermosonic Bonding. Jellison and Wagner [T-23] studied thermosonic bonding characteristics of both thin and thick films. In general, the shear strength improved significantly with increased temperature and more slowly with increased ultrasonic power.

Michaels [T-31] also stated that the strongest TS bonds to aluminum are made at high interface temperatures, 175 °C or higher, and that this is more important than increasing the ultrasonic power. Johnson et al [T-24] stud-

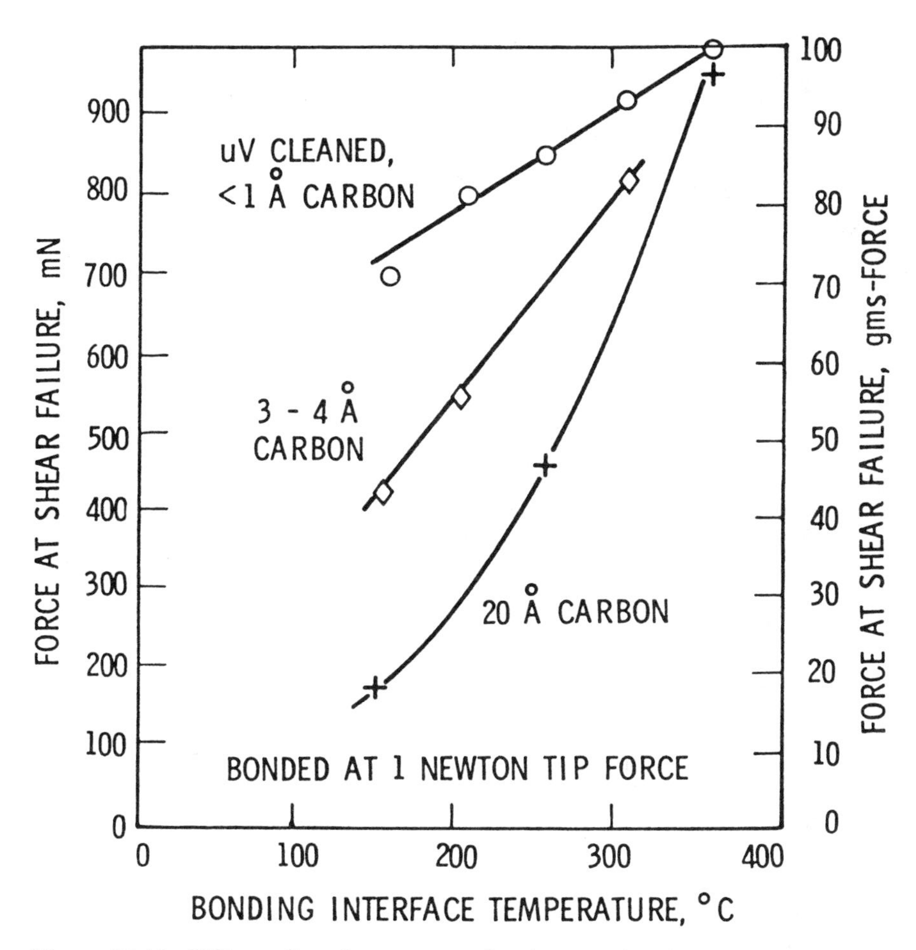

Figure T-18. Effect of surface contamination on the thermocompression bonding of gold, pulsed bonding. Data are for TC ball bonds made with 25-μm (1-mil) diameter gold wire bonded to gold-chromium metallization with a bonding force of 98 gf. The contamination in this figure resulted from exposing substrates to laboratory air for various lengths of time (after Jellison [T-17]).

ied ultrasonic ball bonding to thick film gold at room temperature and found that over a wide range of ultrasonic power the shear strength remained essentially constant after reaching a threshold. The above characteristics of bonding to gold are corroborated by Weiner et al [T-25]. In addition, Weiner et al bonded to thin film aluminum metallizations. Their data are shown in Figure T-19 and are representative of most TS bonding studies

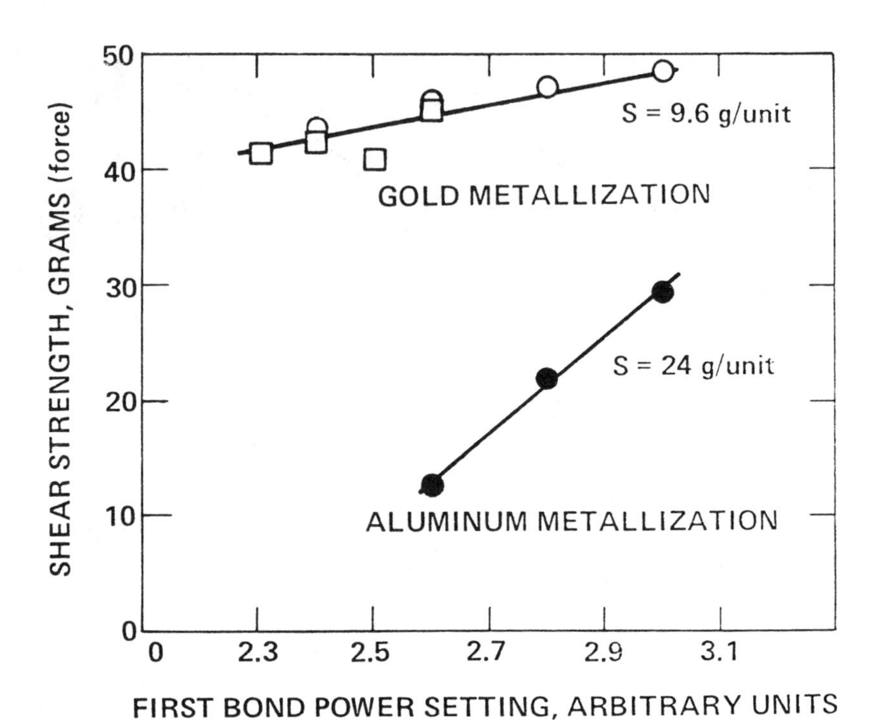

Figure T-19. Effect of ultrasonic power on ball-shear strength. Bonds made with 25-μm (1-mil) diameter gold wire. The pedestal temperature for these measurements was 125 °C and the bonding force was 30 gf (after Weiner et al [T-25]).

on gold but show a higher power dependency for aluminum bonding than reported by others. This may have resulted from using a low, approximately 125 °C, stage temperature.

For TC bonding, it was possible to assemble published data and give typical bonding parameters. This is not possible for TS bonding partly because the published data or parameters do not overlap and partly because the ultrasonic parameter, power, is expressed as dial setting. Only Johnson et al [T-24] gave bonding tool vibration amplitude, but they worked at 25 °C rather than at elevated temperatures, where less power is usually required. The actual ultrasonic energy applied to the bond, for a given dial setting, is dependent upon the specific power supply design as well as the efficiency of the transducer[5] and possibly the characteristics of the capil-

[5]Many ultrasonic power supplies-transducer combinations result in decreasing ultrasonic energy as the bond matures. Such supplies cannot be directly compared with fixed amplitude ultrasonic systems.

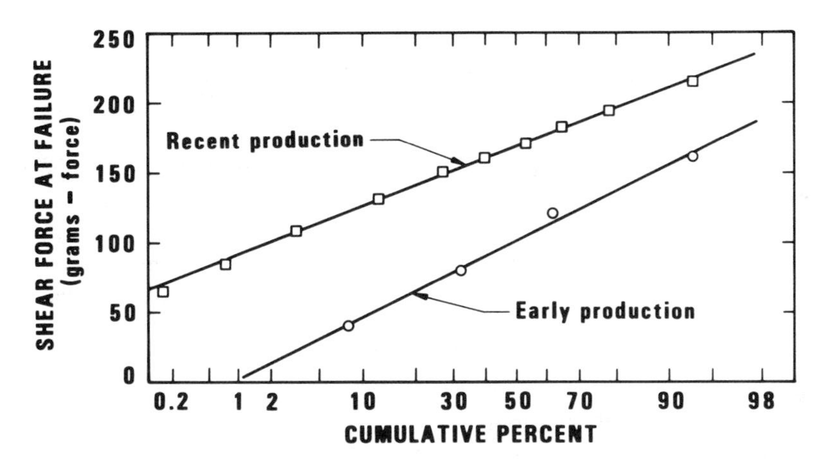

Figure T-20. A comparison of the ball-shear strength distribution of production transistor wire bonds before and after instituting the ball-shear test for setting up bonders and monitoring production. TC bonds were made using 38-μm (1.5-mil) diameter gold wire. (Data replotted from Basseches and D'Altroy [T-34].)

lary. In any event, from references [T-17,23,31], it appears that most elevated temperature TS bonds (which may be over 200 °C) are more nearly thermocompression than ultrasonic bonds, especially if the surface is not contaminated. Since there are no clearly defined optimum parameters, except high temperature, it is extremely important to use the ball-shear test to optimize bonding machine parameters for TS bonding. One can use factorial methods to assist in this optimization. Such factorial methods have been described by Chen [T-32] and Charles [T-33].

Evaluation of Production Bond Quality. The future of the ball-shear test, as with the pull test, lies in the area of production quality control (QC). The first published use of the bond shear test in 1967 was to monitor and control aspects of microelectronic production. Gill and Workman [T-16] used it to monitor adhesion of the then new molybdenum-gold metallization system. Weak adhesion of gold-to-molybdenum was revealed by the gold peeling from the molybdenum during the bond-shear test. This test is still an excellent method to determine the quality of metallization adhesion (see Figure T-14).

Basseches and D'Altroy [T-34] used the ball-shear test to monitor TC bonding for plastic-encapsulated transistors. Their work resulted in reducing the number of hot test failures by a factor of 12. Figure T-20 is their replotted shear test data for 38-μm (1.5-mil) diameter gold wire. This clearly demonstrates the improvement which was directly attributable to instituting the shear test, both to improve bonder setup and to continually sample production. Stafford [T-20] also studied failures in thermally cycled plastic-encapsulated transistors having 25-μm (1-mil) diameter ball bonds that were TS and TC bonded. In his work, as in [T-34], the information from the failed product was used to improve the bonding schedule and resulted in greatly improved yield.

From the last two papers [T-20,34], it is apparent that the high stresses applied to ball bonds in plastic-encapsulated devices during molding and later during thermal cycling make it essential to use the ball-shear test for production bond quality control. When used as a continuing production QC test, the shear test will reveal the effects of recently introduced contamination as well as any variation in the metallization or glassivation removal process. This information can be obtained quickly enough to take corrective action before large amounts of a failed product are made. Shear testing is used for QC by essentially all of the manufacturing organizations that have published papers on some aspect of that subject.

The nondestructive ball shear test has been investigated by Panousis and Fischer [T-21] who found that balls can be prestressed to 75% of their shear strength without significant strength degradation. This test could be used to assure production bond quality as is done with the nondestructive bond pull test. However, if the nondestructive shear test is used, great care is required in positioning the shear tool to avoid damaging the chip passivation and/or metallization. ***This potential danger will probably prevent use of the shear test for nondestructive testing.***

1.4.10 Shear Test for Wedge Bonds

The shear test is clearly useful for evaluating ball bonds, but is it also useful on ultrasonic aluminum wedge bonds? A cooperative experiment between NIST and Sandia Laboratories [T-35] was designed to determine this. The test consisted of three groups of aluminum ultrasonic wedge bonds on a single substrate made at NIST, half of which were pulled to destruction and then the substrate was sent to Sandia for shear testing. The data are shown in Figure T-21.

These results may be understood if the metallurgical nature of the bond is considered. For aluminum ultrasonic bonds, the bond heel becomes metallurgically overworked and weakens as the bond deformation increases, but simultaneously the amount of the welded area increases. The pull test is particularly sensitive to the weakening of the bond heel. Therefore, the pull force decreases as the deformation increases. The shear test, on the other hand, is completely independent of the condition of the heel; it is

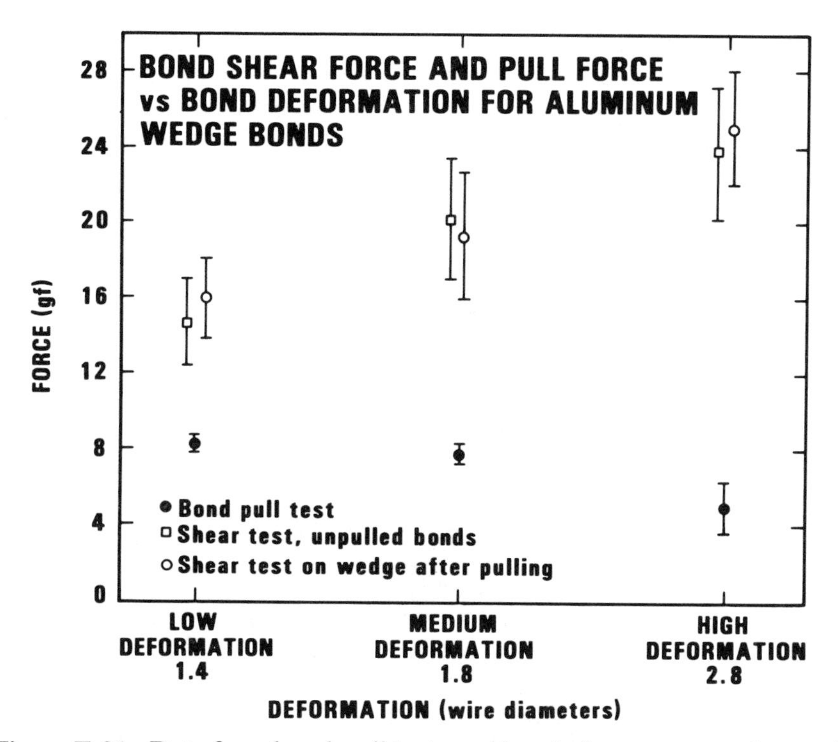

Figure T-21. Data from bond-pull tests and bond-shear tests on ultrasonic aluminum wedge bonds of 25-μm (1-mil) diameter wire on aluminum metallization. The error bars represent one sample standard deviation above and below the mean strength of approximately 20 bonds (after Harman and Cannon [T-4]).

sensitive only to the actual amount of welded area. A high shear value could be obtained from a bond with a cracked or completely broken neck.

From the above, it is apparent that the shear test is not particularly useful for evaluating wedge bonds made from small-diameter wire. However, large aluminum wire wedge-bonds, such as are used in power devices, are usually made with parallel-grooved tools and stand relatively high. Such bonds can be successfully evaluated with a shear tester. A well-bonded aluminum wedge bond from 100-μm (>4-mil) diameter wire on a power device should yield a shear force value in the order of 2 to 4 times the pull force value (depending on the length of the bond), thus greatly increasing the sensitivity for bonding machine setup purposes. Ribbon bonds with a height of 12 μm (approximately 0.5 mil) or higher can also be successfully tested with a shear tester. Optimum shear test values for both ribbon and large-diameter wire can be obtained from the curves of Figure T-15 by correcting for the rectangular bonded area.

1.4.11 Ball-Shear Test Standardization

One reason that the ball-shear test has not been universally adopted is that at present there are no standards for its use. That situation is changing. The ASTM Committee F-1.07 has written a standard ball-shear test method. A round robin has been performed by testing controlled-strength ball bonds at several facilities and using several different types of measurement equipment.

The ASTM standard is in press at this writing. In a short time, that specification will be submitted to government-industry coordination for inclusion as a test method in MIL STD 883. Thus, the ball-shear test has the potential for becoming as universal as the bond pull test.

1.5 THERMAL STRESS TEST FOR WIRE BOND RELIABILITY

Gold bonds on aluminum pads (or the reverse) have long been observed to fail beyond some level of thermal stress (see section on Intermetallics). However, Horsting [T-36] (for Al bonds on gold) and White [T-30] (for gold bonds on Al) found that if no impurities are present in the bond interface, the bond will remain strong even after long thermal stress times. If impurities are in the interface (or the bond is poorly welded), then the bond strength will degrade rapidly during such stress. To reveal potential problems, Horsting applied a stress test consisting of a 390 °C bake for one hour followed by a pull test. If the bonds failed at a low force, by interface separation, the package lot was rejected. Later, Ebel [T-37] introduced a bake schedule as a screening procedure to reveal similar potential bond failures for hybrids. Most recently, MIL STD 883, Method 5008 [T-38], has specified a similar, though less severe, test. The time, temperature, and other conditions of these various stress tests are given in Table T-3.

TABLE T-3
VARIOUS THERMAL STRESS TESTS FOR Au-Al BOND RELIABILITY ASSESSMENT

Time (hrs)	*Temperature °C*	*Pull Test Failure Criteria*	*Reference*
1	390°	Weak or "lifts"	Horsting [T-36]
1 4 24 200 3000	350 300 250 200 150	less than one half the minimum acceptable post seal pull strength of (MIL STD 883, Method 2011)	Ebel [T-37]
1	300 °C	<1 gm for 1-mil 25-µm wire <0.5 gm for smaller wire	MIL STD 883 [T-38]

1.6 APPENDIX T-1

THE SHELF-LIFE AGING OF BONDING WIRE

Small organizations and those that only occasionally use a particular type of wire are often concerned about the aging properties of bonding wire. Some discard wire after an arbitrary period, such as three or six months, rather than risk a change in its metallurgical properties which might affect bonding. Recently, ASTM[1] (Committee F-1.07) had several U.S. wire manufacturers study the actual aging properties of various 25-μm (1-mil) diameter bonding wires. These were stored on 2-in spools at 73 ±3 °F for two years and tested periodically.

In general, the breaking load of hard, as-drawn wire decreased rapidly (from 5 to 15%) within six weeks after manufacture and continued to decrease over the two-year period. The exception was hard aluminum, 1% magnesium wire, which was essentially unchanged over the two-year test period. All gold and aluminum stress-relieved and annealed wires stayed within their breaking load specification for the entire two-year test period. The elongation characteristics were more ambiguous than the breaking load, changing upward or downward but within the specification

[1]American Society for Testing Materials, 1916 Race Street, Philadelphia, PA 19103.

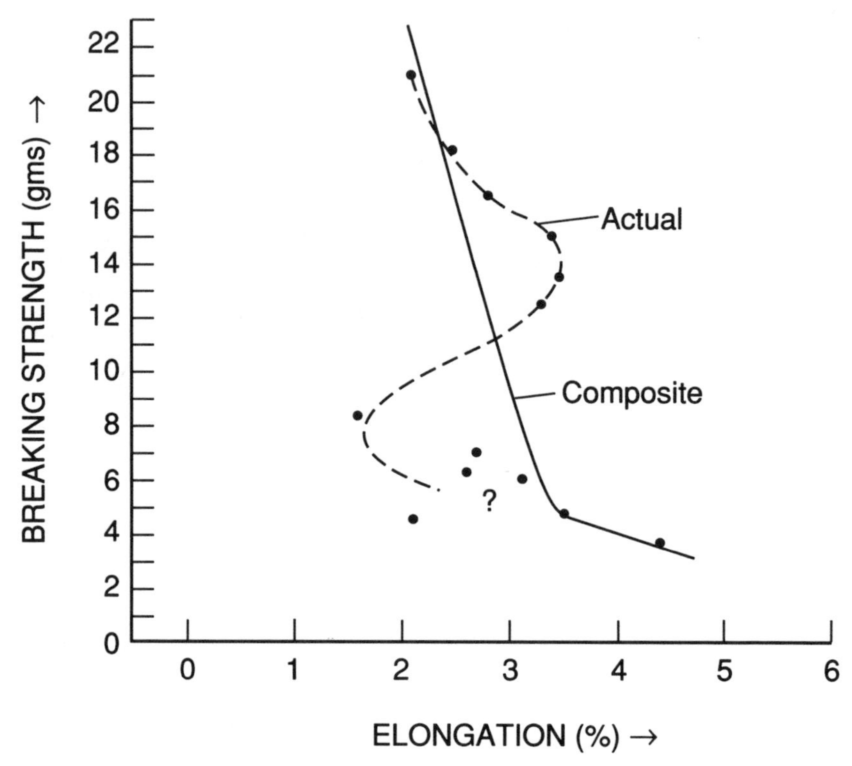

Figure TA-1. The breaking strength versus elongation for 250-μm (10-mil) diameter aluminum wires. Most large bonding wire is used in the high elongation (>15%) condition (American Fine Wire Company [T-40]).

extremes, and generally recovering to the median by the end of the test. The data were compiled and published in ASTM Standards F487 (for various aluminum alloys, (1% Si and 1% Mg) and F72 (for various gold alloys, Be, Cu, or Ag doped), and the reader is directed to these ASTM standards for detailed data.

The conclusion drawn from this study is that, in general, small-diameter wire (except hard, as drawn) can be used for up to two years with only minimal change in its breaking load, although its elongation may vary over its entire specified range. The *caveat* is that the wire must be stored at approximately constant room temperature and exposure to direct sunlight, drafts from an open door, or possible heat sources ***must be avoided.***

No similar aging studies have been done with larger-diameter bonding wire > 25-μm (1-mil). Generally, most large-diameter aluminum wire used for power devices is 99.99% aluminum or contains 0.5% magnesium, and

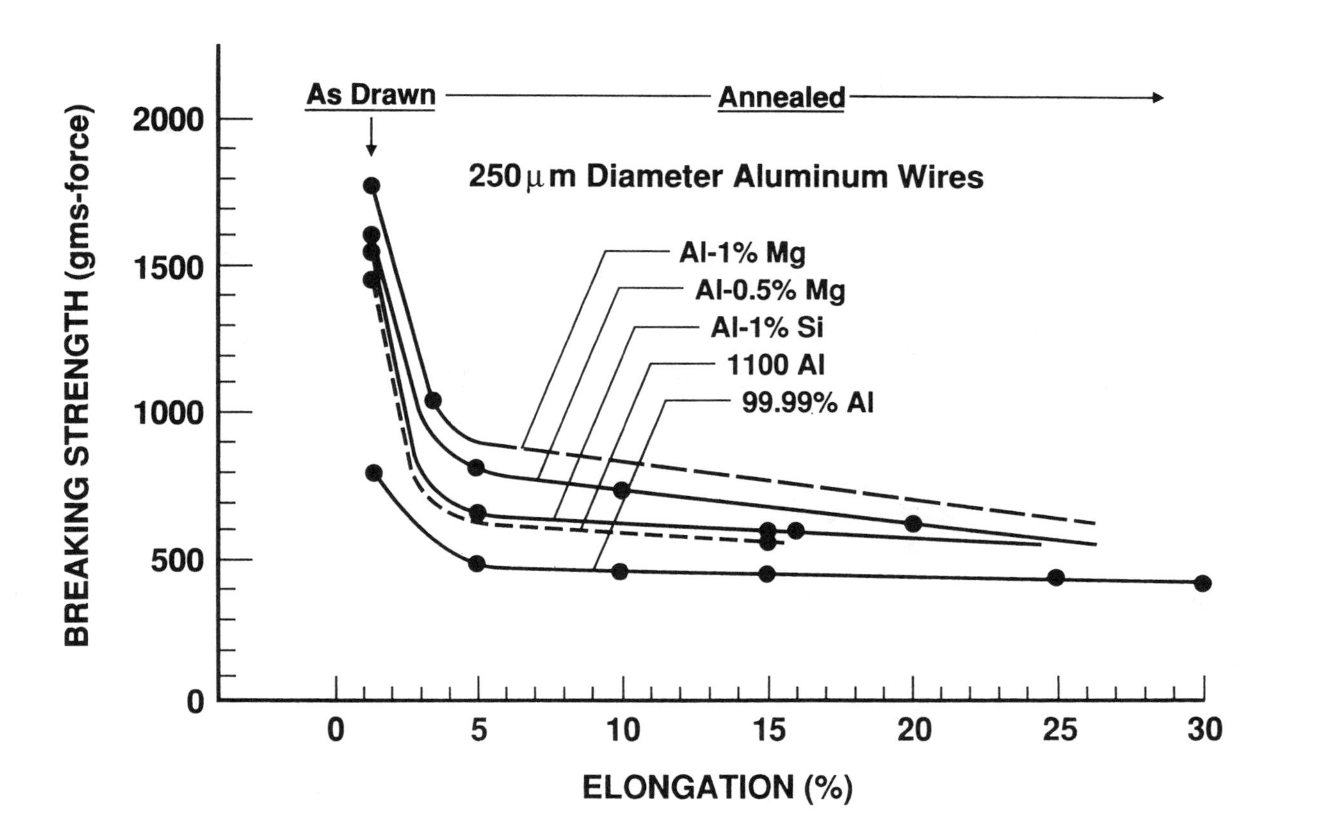

Figure TA-2. The breaking strength versus elongation for 25-μm (1-mil) diameter 1% Si aluminum wire. The variation in elongation in the intermediate range results in drift of that parameter during aging [T-40].

is used in the fully annealed condition (>15% elongation). Based on the above tests for small wire, it can be assumed that the breaking load is not apt to change over a two-year period, but the elongation, which is generally >15%, could change somewhat. This will generally have little effect on bonding such wire ultrasonically. An example of typical breaking loads and elongations for large wires is given in Figure TA-1, and for Al 1% Si 25-μm (1-mil) diameter wire in Figure TA-2. The small silicon-doped wire is quite different from the large wire due to the combination of silicon precipitates and grain boundary growth being limited in size by the wire diameter. The exact shape of the curve will vary with annealing temperature, cooling time, and individual manufacturer's process. The scatter in elongation accounts for the changes with aging discussed above.

1.7 REFERENCES

T-1. Polcari, S. M., and Bowe, J. J., Evaluation of Non-Destructive Tensile Testing, Report No. DOT-TSC-NASA-71-10, 1-46 (June 1971).

T-2. Schafft, H. A., Testing and Fabrication of Wire-Bond Electrical Connections—A Comprehensive Survey, National Bureau of Standards Tech. Note 726, (1972).

T-3. Albers, J. H., Editor, Semiconductor Measurement Technology. The Destructive Bond Pull Test, NBS Spec. Publ. 400-18 (February 1976).

T-4. Harman, G. G., and Cannon, C. A., The Microelectronic Wire Bond Pull Test, How to Use It, How to Abuse It, IEEE Trans. on Components, Hybrids, and Manufacturing Technology CHMT-1, pp. 203-210, (September 1978).

T-5. Harman, G. G., Editor, Semiconductor Measurement Technology. Microelectronic Ultrasonic Bonding, NBS Spec. Publ. 400-2 (January 1974).

T-6. Ang, C. Y., Eisenberg, P. H., and Mattraw, H. C., Physics of Control of Electronic Devices, Proc. 1969 Annual Symposium on Reliability, Chicago, Illinois, pp. 73-85 (January 1969).

T-7. Slemmons, J. W., The Microworld of Joining Technology, The American Welding Society 50th Annual Meeting, Proceedings, Philadelphia, Pennsylvania, pp. 1-48 (April 1969).

T-8. Bertin, A. P., Development of Microcircuit Bond-Pull Screening Techniques, Final Technical Report RADC-TR-73-123 (April 1973), AD762-333.

T-9. Harman, G. G., A Metallurgical Basis for the Non-Destructive Wire-Bond Pull-Test, 12th Annual Proceedings IEEE Reliability Physics, Las Vegas, Nevada, April 2-4, 1974, pp. 205-210.

T-10. Plumbridge, W. J., and Ryder, D. A., The Metallography of Fatigue, Metallurgical Reviews 14-15, p. 129 (1970); and Alden, T. H., and Backofen, W. A., The Formation of Fatigue Cracks in Aluminum Single Crystals, Acta Metallurgica 9, pp. 352-366 (1961).

T-11. Roddy, J., Spann, N., and Seese, P., IEEE Trans. on Components, Hybrids, and Manufacturing Technology 1, pp. 228-236 (1978).

T-12. Blazek, R. S., Development of Nondestructive Pull Test Requirements for Gold Wires on Multilayer Thick Film Hybrid Microcircuits, IEEE Trans. on Components, Hybrids, and Manufacturing Technology CHMT-6, pp. 503-509 (December 1983).

T-13. Harman, G. G., and Leedy, K. O., An Experimental Model of the Microelectronic Ultrasonic Wire Bonding Mechanism, 10th Annual Proc. IEEE Reliability Physics, Las Vegas, Nevada, April 5-7, 1972, pp. 49-56, and Harman, G. G., Acoustic Emission Monitored Tests for TAB Inner Lead Bond Quality, IEEE Trans. on Components, Hybrids, and Manufacturing Technology CHMT-5, pp. 445-453 (1982).

T-14. Harman, G. G., The Microelectronic Ball-Bond Shear Test - A Critical Review and Comprehensive Guide to Its Use, The International Journal for Hybrid Microelectronics 6, pp. 127-141 (1983).

T-15. Arleth, J. A., and Demenus, R. D., A New Test for Thermocompression Microbonds, Electronic Products 9, pp. 92-94, May 1967.

T-16. Gill, W., and Workman, W., Reliability Screening Procedures for Integrated Circuits, Physics of Failure in Electronics 5, RADC Series in Reliability; T. S. Shilliday and J. Vaccaro, Eds., June 1967, pp. 101-142.

T-17. Jellison, J. L., Effect of Surface Contamination on the Thermocompression Bondability of Gold, Proc. 28th IEEE Electronic Components Conference, Washington, DC, May 11-12, 1975, pp. 271-277; also IEEE Trans. Parts, Hybrids, and Packaging PHP-11, pp. 206-211 (1975).

T-18. Jellison, J. L., Kinetics of Thermocompression Bonding to Organic Contaminated Gold Surfaces, Proc. 29th IEEE Electronics Components Conf., San Francisco, California, April 26-28, 1976, pp. 92-97; also IEEE Trans. Parts, Hybrids, and Packaging PHP-13, pp. 132-137 (1977).

T-19. Shimada, W., Kondo, T., Sakane, H., Banjo, T., and Nakagawa, K., Thermo-Compression Bonding of Au-Al System in Semiconductor IC Assembly Process, Proc. Int. Conf. on Soldering, Brazing, and Welding in Electronics, DVS, Munich, November 25-26, 1976, pp. 127-132.
T-20. Stafford, J. W., Reliability Implications of Destructive Gold-Wire Bond Pull and Ball Bond Shear Testing, Semiconductor International 5, pp. 83-90 (May 1982).
T-21. Panousis, N. T., and Fischer, M. K. W., Non-Destructive Shear Testing of Ball Bonds, Intl. J. for Hybrid Microelectronics 6, pp. 142-146 (1983).
T-22. Charles, Jr., H. K., Johns Hopkins University, Applied Physics Laboratory, private communication.
T-23. Jellison, J. L., and Wagner, J. A., The Role of Surface Contaminants in the Deformation Welding of Gold to Thick and Thin Films, Proc. 29th IEEE Electronics Components Conference, Cherry Hill, New Jersey, May 14-16, 1979, pp. 336-345.
T-24. Johnson, K. I., Scott, M. H., and Edson, D. A., Ultrasonic Wire Welding, Part II, Ball-Wedge Wire Welding, Solid State Technology 20, 91-95 (1977).
T-25. Weiner, J. A., Clatterbaugh, G. V., Charles, Jr., H. K., and Romenesko, B. M., Gold Ball Bond Shear Strength, Effects of Cleaning, Metallization, and Bonding Parameters, Proc. 33rd IEEE Electronics Components Conf., Orlando, Florida, May 16-18, 1983, pp. 208-220.
T-26. Philofsky, E., Purple Plague Revisited, Solid-State Electron. 13, pp. 1391-1399 (1970).
T-27. Devaney, J. R., and Devaney, R. M., Thermosonic Ball Bond Evaluation by a Bond Pluck Test, Proceedings 1984 ISTFA Conference, Los Angeles, California, pp. 237-242.
T-28. Kashiwabara, M., and Hattori, S., Formation of Al-Au Intermetallic Compounds and Resistance Increase for Ultrasonic Al Wire Bonding, Review of the Electrical Communication Laboratory (NTT) 17, pp. 1001-1013 (September 1969).
T-29. Clatterbaugh, G. V., Weiner, J. A., and Charles, Jr., H. K., Gold-Aluminum Intermetallics: Ball Bond Shear Testing and Thin Film Reaction Couples, IEEE Trans. on Components, Hybrids, and Manufacturing Technology CHMT-7 (1984), pp. 349-356. Also see earlier publications by Charles and Clatterbaugh, International Journal for Hybrid Microelectronics 6, pp. 171-186 (1983).
T-30. White, M. L., Serpiello, J. W., Stringy, K. M., and Rosenzweig, W., The Use of Silicone RTV Rubber for Alpha Particle Protection

on Silicon Integrated Circuits, 19th Annual Proc., Reliability Physics 1981, Orlando, Florida, April 7-9, 1981, pp. 43-47.

T-31. Michaels, D., Burroughs Corp. (UNISYS), comments made at ASTM Committee F-1 Meeting, San Diego, California, February 1-2, 1983.

T-32. Chen, Y. S., and Fatemi, H., Gold Wire Bonding Evaluation By Fractional Factorial Designed Experiment, The International Journal for Hybrid Microelectronics 10, Number 3, 1987, pp. 1-7.

T-33. Charles, Jr., H. K., and Clatterbaugh, G. V., Ball Bond Shearing - A Complement to the Wire Bond Pull Test, Intl. J. Hybrid Microelectronics 6, No. 1, pp. 171-186, (October 1983).

T-34. Basseches, H., and D'Altroy, F., Shear Mode Wire Failures in Plastic-Encapsulated Transistors, IEEE Trans. Components, Hybrids, Manufacturing Technology CHMT-1, pp. 143-147 (1978).

T-35. NBS Special Publication 400-19, Semiconductor Measurement Technology Quarterly Report, January 1 to June 30, 1975, W. M. Bullis, ed. (April 1976), p. 51. The NBS portion of the work was done by G. G. Harman and C. A. Cannon and the Sandia portion by Wayne Vine.

T-36. Horsting, C. W., Purple Plague and Gold Purity, 10th Annual Proc. IEEE Reliability Physics Symposium, Las Vegas, Nevada, pp. 155-158 (1972).

T-37. Ebel, G. H., Failure Analysis Techniques Applied in Resolving Hybrid Microcircuit Reliability Problems, 15th Annual Proceedings Reliability Physics, Las Vegas, Nevada, April 12-14, 1977, pp. 70-81.

T-38. MIL STD 883 Method 5008, Test Methods and Procedures for Microelectronics.

T-39. Schultz, G. and Chan, K., A Quantitive Evaluation of Compound Ball Bonds, 1988 Proc. Intl. Symposium on Microelectronics (ISHM), Seattle, Washington, October 12-19, 1988, pp. 238-245.

T-40. Douglas, P., American Fine Wire Company, private communication. For a discussion of some small diameter bonding wire characteristics see also, New Bonding Wire Developments, Microelectronic Packaging Technology: Materials and Processes, Proc. of the 2nd ASM International Electronic Materials and Processing Congress, Philadelphia, Pennsylvania, April 24-28, 1989, pp. 8902-8908.

CHAPTER 2

INTERMETALLIC COMPOUND AND OTHER BIMETALLIC INTERFACE RELATED BOND REACTIONS

2.1 GOLD-ALUMINUM INTERMETALLIC COMPOUND FORMATION AND CLASSICAL FAILURES

2.1.1 Introduction

Gold-aluminum intermetallic compound formation and associated Kirkendall void formation have resulted in more documented bond failures than any other problem over the years. There have been hundreds of papers on this subject, and this section can only present an overview of them. Modern packaging and device environments do not (or need not) involve the high temperatures required to produce these classical failures. Most present-day Au-Al related failures are more properly referred to as impurity-driven or corrosion reactions. These are discussed in the following section, but an understanding of the classical failures is essential to comprehend the present day failures.

The most definitive work on Au-Al compounds, oriented towards microelectronics, was done by Philofsky [I-1 to 3] and those interested in details are referred to his publications. ***The compounds (often called purple plague[1] because of the appearance of the $AuAl_2$ compound) will occur between gold wire bonded to aluminum metallization, and vice-versa, if the temperature-time product is large enough.*** Such compounds will begin to form during the process of thermocompression or thermosonic bonding. They may also occur during qualification screens, plastic encapsulation cure, or at any time when high temperatures are encountered during the life of a device. A few monolayers of such compounds will even form at room temperature.

2.1.2 Intermetallic Compound Formation

There are five intermetallic compounds as shown in the phase diagram [I-4] of Figure I-1. These compounds, as many other intermetallic compounds, are colored, with $AuAl_2$ being purple and the rest being tan or white as indicated in Figure I-2. Since the phases are usually mixed, the observed color is often gray, brown, or black. The aluminum-rich $AuAl_2$ has a high melting point and therefore is relatively stable. In general, however, under continued thermal exposure, diffusion continues (especially through the low melting point compounds) until the gold or the aluminum is consumed. After this, there can be a rearrangement towards the excess metal-rich compounds (Au-rich in the case of a ball bond on thin Al metallization). But, in general, the reaction slows as was demonstrated by the ball-shear test shown in Figure T-17 of the testing section.

Observations suggest that the initial growth rate of the intermetallic alloys usually follow a parabolic relationship:

$$\mathbf{x} = \mathbf{K}\mathbf{t}^{1/2} \qquad (\mathrm{I}-1)$$

where x is the intermetallic layer thickness, t is the time, and **K** is the rate constant and:

$$\mathbf{K} = \mathrm{C} \cdot \exp\left[\frac{-\mathrm{E}}{\mathrm{kT}}\right] \qquad (\mathrm{I}-2)$$

where C is a constant, E is the activation energy for layer growth (in electron volts), k is the Boltzman constant, and T is the absolute temper-

[1]Legend has it that in the early 1960's an anonymous engineer was looking through a microscope at a ball bond on a failed transistor. He saw a purple ring around the bond and exclaimed "My God, the transistors have caught the purple plague." The name stuck.

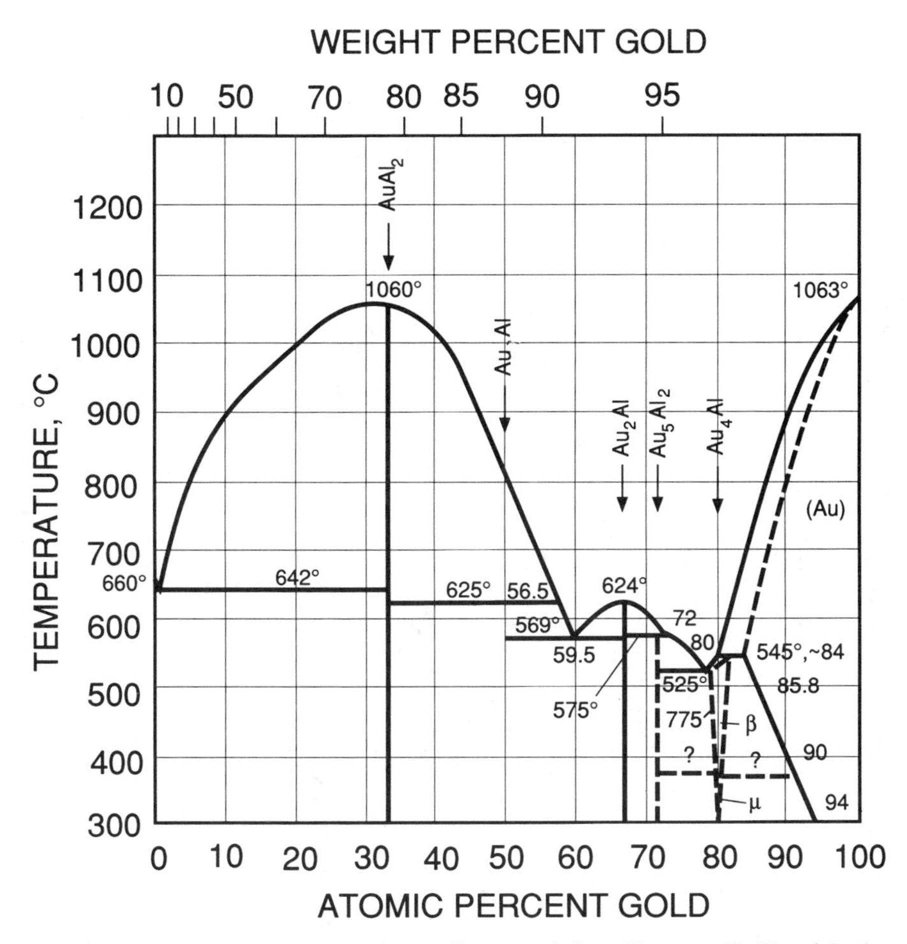

Figure I-1. Aluminum-gold phase diagram (after Hansen [I-4]) with the five Au-Al intermetallic compounds indicated.

ature (in kelvins).[2] The value of **K** changes for each intermetallic phase, and is also dependent upon the neighboring phases, which supply additional Au and Al for continued compound formation. Because of this, Philofsky lists nine different rate constants for the five Au-Al compounds. Figure I-2 shows the relative rates of intermetallic formation. From this, it is apparent that Au_5Al_2 grows much faster than the other phases.

[2]In chemical or metallurgical literature, one often sees the equation written as: $\mathbf{K} = C\exp[-Q/RT]$ where Q is the activation energy in kilocalories/mole (1 eV ≈ 23 K Cal/mole), R is the gas constant (1.98) and T = temperature in kelvin.

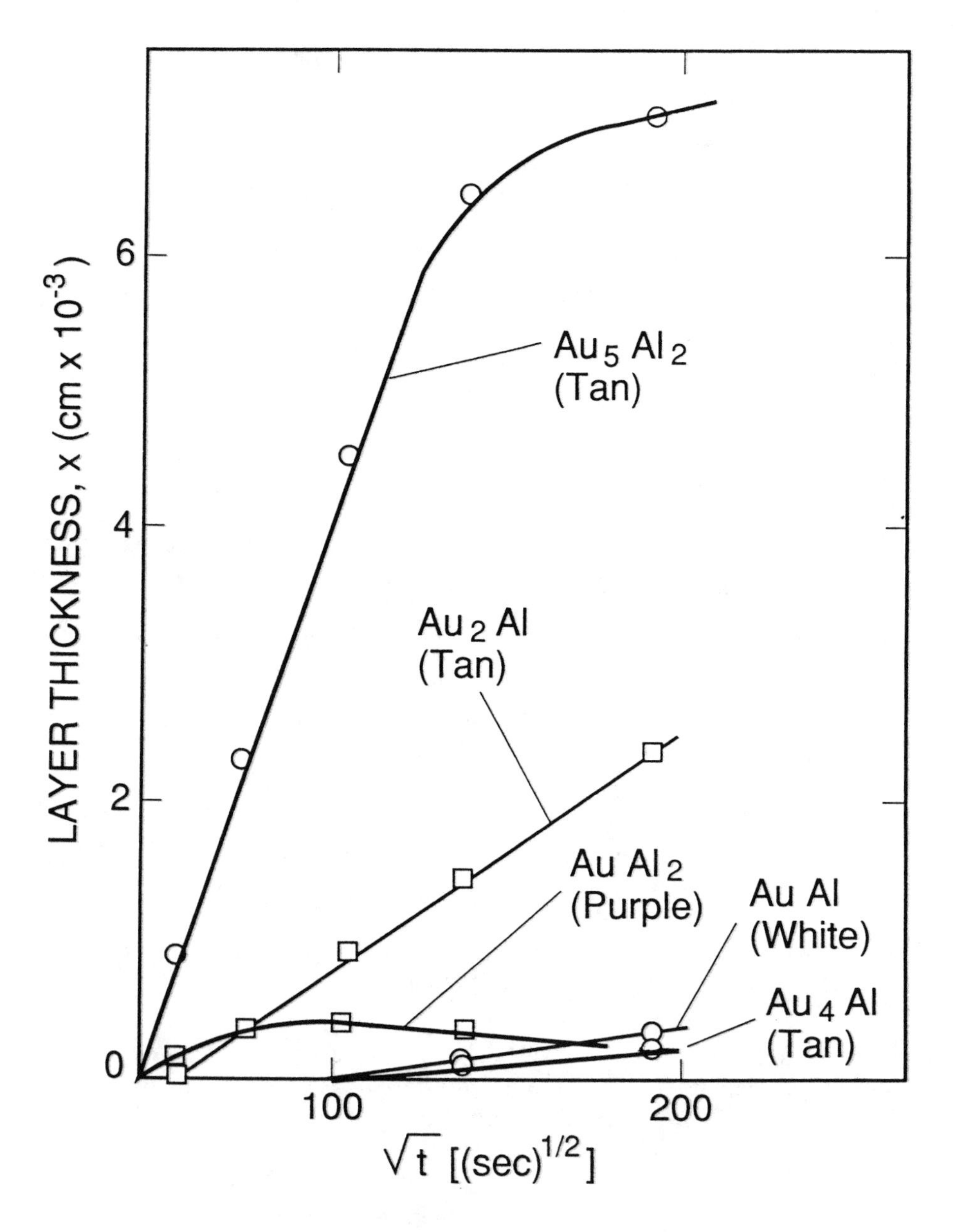

Figure I-2. Layer thickness of the five intermetallic phases versus the square root of time at 400 °C. Data were obtained on large butt-welded couples having unlimited and equal supplies of Au and Al (after Philofsky [I-1]).

The rate of diffusion of one metal into the other (or into itself) is dependent on defects in the crystal lattice. Defects can be vacancies, dislocations, and grain boundaries. During diffusion, one atom moves into an empty lattice position (vacancy) and another atom moves into the empty position of the first. Grain boundaries and surfaces, because the lattice has more open structures, have many vacancies and can speed up diffusion rates by orders of magnitude compared to diffusion in the bulk. Poorly welded bonds consist of numerous isolated microwelds which contain large surface areas to volume ratios, as well as mechanical stresses which result in numerous lattice defects. Thick film metallizations also contain many grain boundaries, stresses, and impurities, all of which result in lattice defects. Thus, it is not surprising that poorly welded bonds or aluminum wire bonds to thick film gold fails rapidly; see Appendix I-1.

A generic activation energy (E) (one that combines the effects of all five compounds and/or Kirkendall voids) for various bond failures is often measured by workers. As a result, the literature abounds with different values of (E) for various properties thought to be related to intermetallic-compound formation. Gerling [I-5] published an extensive compilation of reported activation energies for various types of Au-Al wire-bond failures (see Table I-1). It is not possible to explain the wide variation in values except that the measurements were made by different methods and were not necessarily related to the same type or definition of failure. Different metallurgical couples were used (Al wire to various gold films, Au wire to various IC metallizations) resulting in Al- or Au-rich couples. Also, some activation energies for bond failure may have resulted from impurities in the interface or the bonds may have been poorly welded (see Appendix I-1).

It should be noted that the specific intermetallic compounds in a bond-interface area are related to the relative amounts of gold and aluminum present. In addition, some compounds may be absent because of a low nucleation probability or they may grow so slowly that they are not observed. Figure I-3 gives the compounds observed to form in gold-rich and aluminum-rich areas, and in areas with gold and aluminum in equal amounts. In addition to this, silicon may form ternary compounds with gold and aluminum but, as shown by Philofsky, these are no more detrimental to bond quality than the pure gold-aluminum compounds by themselves.

These intermetallic compounds are ***not*** the normal cause of failure. They are mechanically strong, although brittle, and electrically conductive. Bond failures result from the formation of Kirkendall voids as well as from the susceptibility of gold-aluminum couples to degradation by impurities and to corrosion. The latter two causes are extensively discussed in following sections. Kirkendall voids form when either the aluminum or gold diffuses

TABLE I-1
VARIOUS THERMAL ACTIVATION ENERGIES FOR BOND FAILURES AND GROWTH OF Au-Al COMPOUNDS (AFTER GERLING [I-5] WITH ADDITIONS)

Ref. [I]	Specimen	Observed Quantity	Thermal Activation
6	Au-Al-films	Au-Al-growth rate	1 eV
7	Au-Al-films	sheet resistance	1 eV
8	Au-Al-wire couples	Au-Al-growth rate	0.78 eV
1	Au-Al-wire couples	Au-Al-growth rate	0.69 eV
2	Au-Al-wire couples	mechanical degradation	1 eV
9	Au-wire, Al-film	Au-Al-growth rate	0.88 eV
10	Au-wire, Al-film 1.4 μm on Ta	contact resistance, $\Delta R = 50\%$	0.55 eV
11	Au-wire, Al-film		
	<0.3 μm	contact resistance, $\Delta R =$ 1 ohm	0.7 eV
	0.5, 1 μm	contact resistance, $\Delta R =$ 1 ohm	0.9 eV
		pull strength (time to failure)	0.2 eV
5	Au-balls, Al-films	resistance (peripheral voiding)	
	1 μm, Al-Si		0.9 eV
	1.3 μm, Al		≥0.8 eV
	2.5 μm, Al		0.6 eV
12	Au-balls, Al-film	ball shear strength	0.56 eV

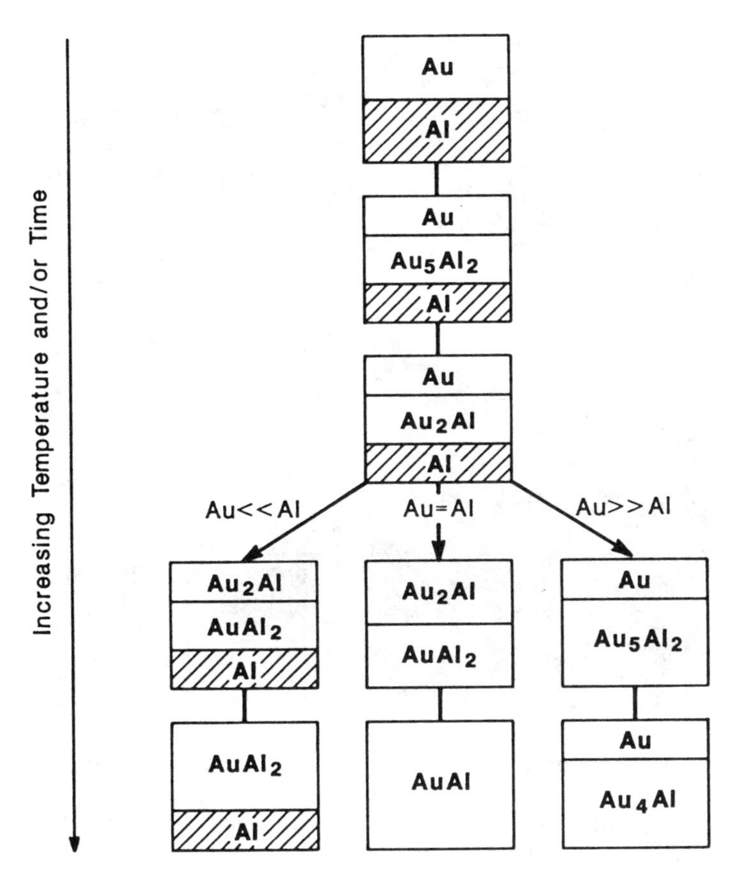

Figure I-3. Schematic representation of compound formation in gold-aluminum thin-film systems. The identity of the final compounds is determined by the annealing temperature and by the proportions of the starting materials. The final compounds result from the reaction being driven to completion (stability) with one component being completely consumed. This occurs only after long times at high temperatures (after G. Majni et al [I-13]).

out of one region faster than it can diffuse in from the other side of that region. Vacancies pile up and condense to form voids, normally on the gold-rich side along the Au_5Al_2-to-Au interface. The rates of diffusion vary with temperature, and with different phases, and are dependent upon the adjacent phases, as well as the number of vacancies in the original metals.

Classical Kirkendall voids typically require bake times greater than an hour at temperatures greater than 300 °C to occur on the gold-rich side

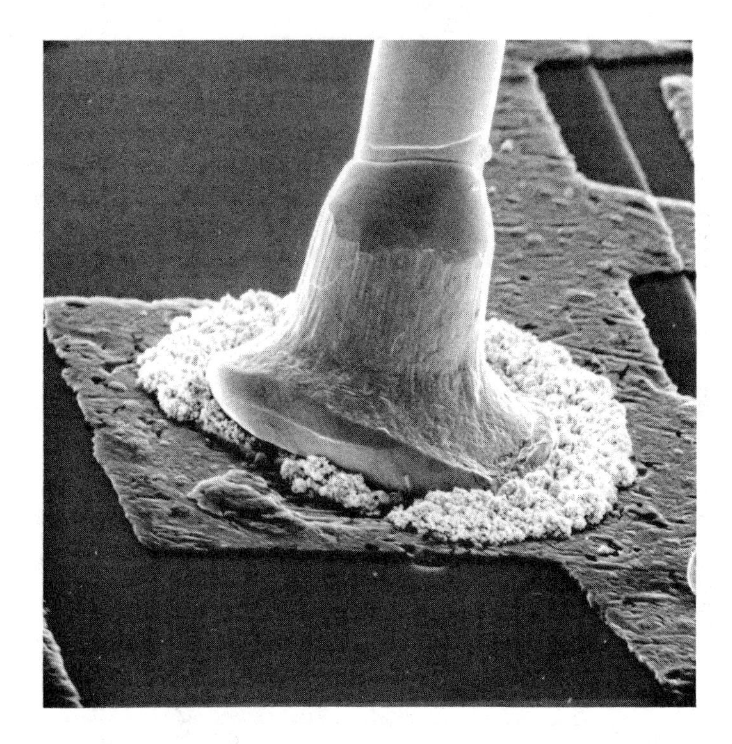

Figure I-4. A SEM photograph of gold-aluminum intermetallic compound formation (white and fluffy) around the perimeter of the bond and under the grossly deformed ball. Even with its poor appearance, the bond was strong and electrically conductive.

(Au_5Al_2), or greater than 400 °C on the aluminum-rich side ($AuAl_2$), or much longer times at lower temperatures [I-3]. Such temperatures and times are seldom reached during modern bonding or modern device and systems packaging. Thus, it is rare that well-made bonds on integrated circuits actually fail due to the formation of classical Kirkendall voids. However, the failures resulting from impurities (see section 2.2), poor welding (see Appendix I-1), or hydrogen in a plated gold layer (see the section on plating) can appear to have resulted from classical Kirkendall voiding, and thus, it is essential to understand the classical failure modes.

2.1.3 Classical Gold-Aluminum Compound Failure Modes

An example of Au-Al compound formation is shown in Figure I-4. Here, a poorly formed ball-bond was subjected to high temperature, and the reaction generated considerable intermetallic compound. However, this

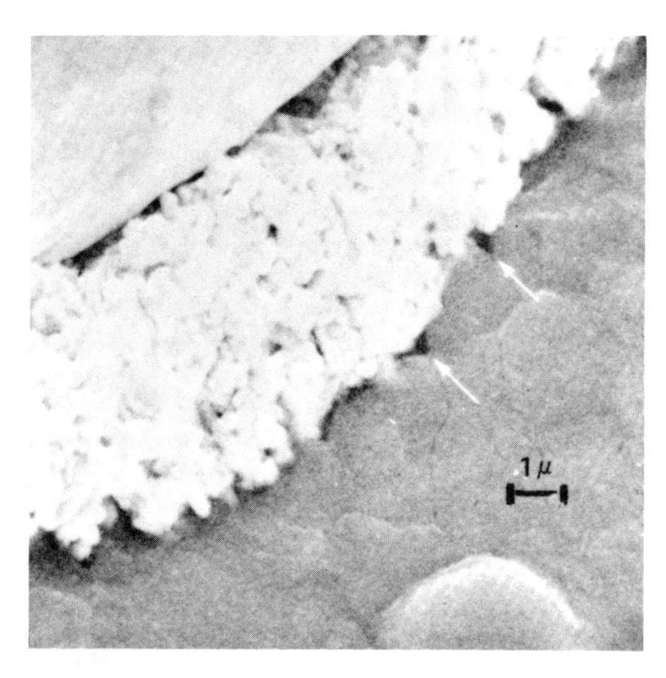

Figure I-5. A SEM photograph of a gold wedge bond to aluminum metallization, aged at 450 °C for 10 minutes, illustrating the voids, indicated by arrows, which form around the periphery of the bond (after Philofsky [I-1]).

bond was both electrically conductive and mechanically strong. Thus, the presence of these compounds will not necessarily cause bonds to fail.

There are three classical bond-failure modes associated with the formation of Au-Al intermetallics. In the first, the bond may be mechanically strong, but can have a high-electrical resistance or may even be open-circuited. In this case, which typically occurs with gold wire-bonded to thin ($\approx$1-μm) aluminum metallization, voids form around the bond periphery limiting the available electrical conduction paths. These voids are indicated by the arrows in Figure I-5 and in the higher magnification photograph of Figure I-6. In the second type of failure, the voids lie beneath the bond, as illustrated in the metallurgical cross section of an aluminum-wire bond over a gold plating, shown in Figure I-7. In this case, the bond can fail due to mechanical weakness. An equivalent section of a gold ball, bonded to aluminum IC metallization is shown in Figure I-8. Note that the intermetallic compound, as well as the voids, rises up just inside the bond perimeter. (No welding takes place on the perimeter.) The diffusion rate

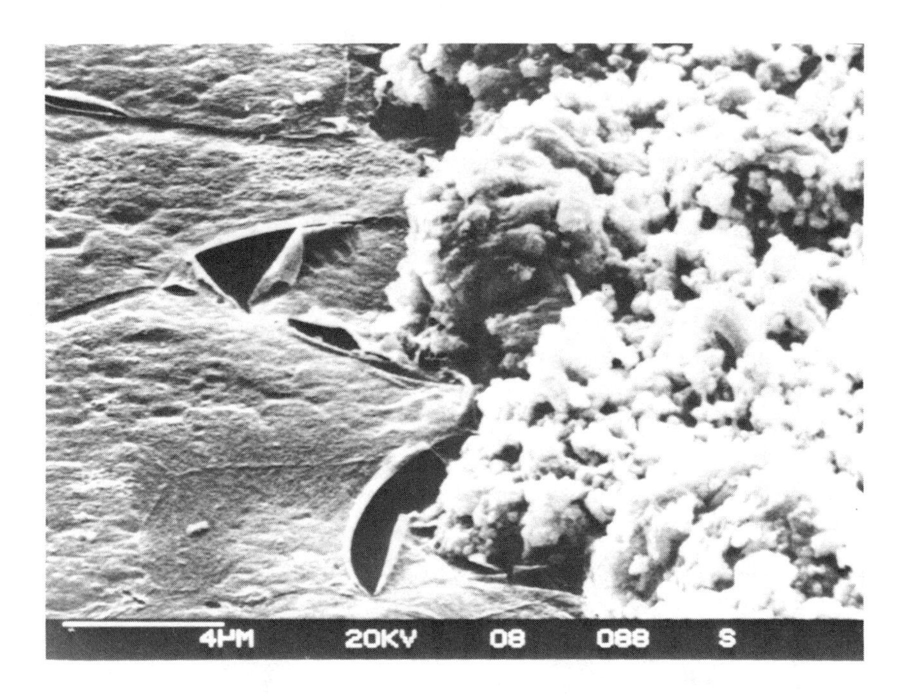

Figure I-6. A closeup of Kirkendall voids undercutting the Al_2O_3 layer on the surface (after Gerling [I-5]).

inside the perimeter is enhanced due to the large number of defects left in this region during the stress and deformation of bonding. It is also the first gold area to be reached by in-diffusing aluminum from the outer bond pad. Gold-aluminum intermetallic compounds are stronger than the pure metals [I-1] when they are void free; however, they are also more brittle [I-2]. Thus, if a wire-bond system contains intermetallics, that system is far more susceptible to brittle fracture during thermal-cycle-induced flexure than gold or aluminum wires alone. An example of a plagued, fatigued, gold wedge-bond which had been cycled only 10 times is given in Figure I-9. In addition to brittleness, the ***growth*** of intermetallic potential compounds is enhanced by temperature cycles. Thus, it is important to be aware of this problem in devices that are likely to be temperature-cycled.

Philofsky [I-2] published metallurgical design limits for avoiding bond failures due to the formation of intermetallic compounds. A condensed version of his diagnostic tables is given in Table I-2. Philofsky, as well as Kashiwabara and Hattori [I-8], reported that thinner metallization will limit Kirkendall voiding by restricting the availability of one of the inter-

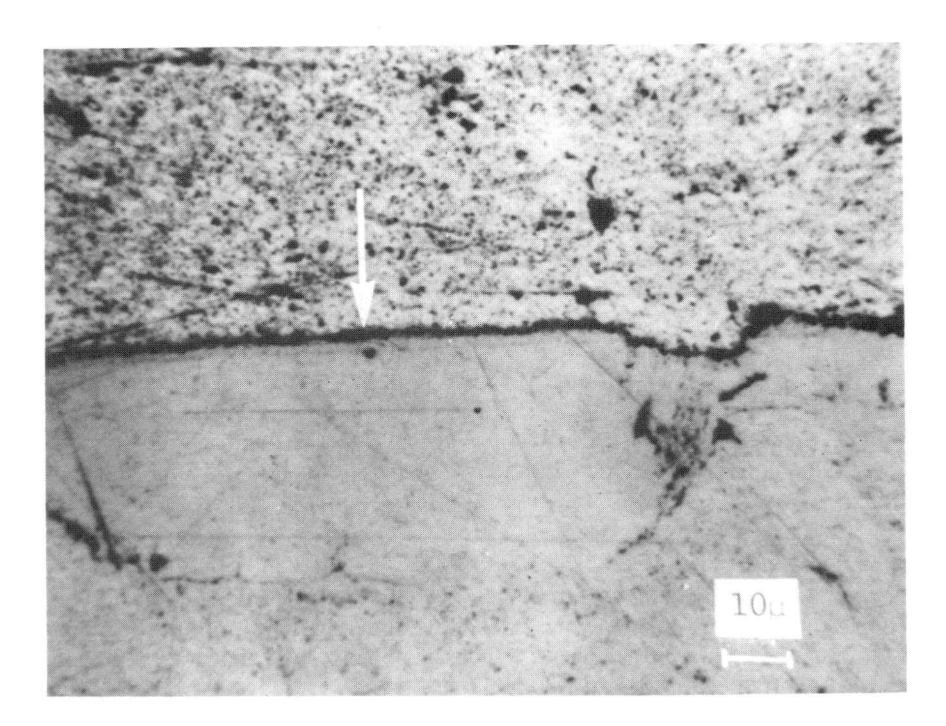

Figure I-7. Al bond on Au aged at 460 °C for 100 minutes. The arrow points to the continuous line of Kirkendall voids that would cause a weak or zero pull-strength bond (after Philofsky [I-2]).

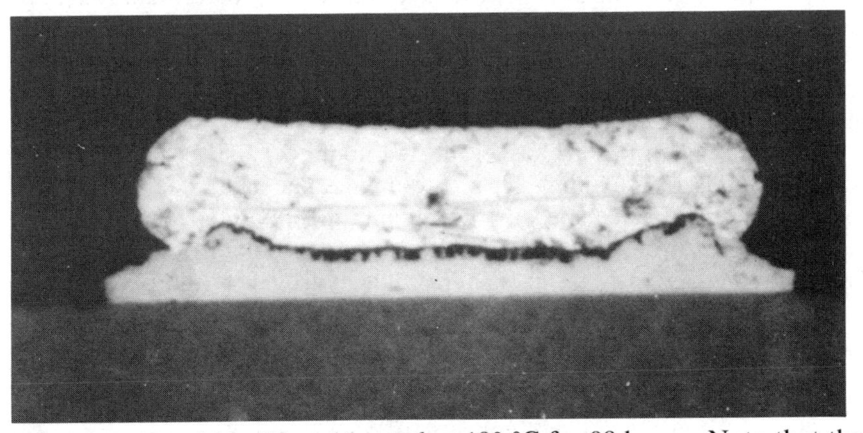

Figure I-8. A gold ball bond heated at 180 °C for 98 hours. Note that the line of voids and the intermetallic compound both rise up just inside the perimeter of the bond. No welding takes place at the perimeter.

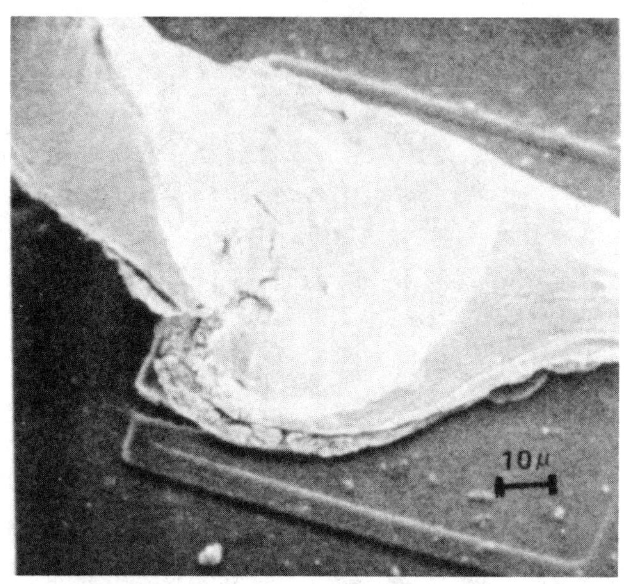

Figure I-9. A SEM photograph of a gold wedge bond to aluminum showing a fatigue crack which formed at the heel after only 10 cycles between −65 and 150 °C (from Philofsky [I-1]).

metallic components. The latter authors also found experimentally that ***no*** intermetallic failures would occur if the ratio of the width of the aluminum-wedge bond to the thickness of the gold film were greater than four, and the storage temperature were less that 350 °C. Larger ratios were needed to avoid failure if higher temperatures were encountered. The width effect was well documented in their paper, but no clear explanation for it was given.

Shih and Ficalora [I-14] have reported inhibiting the formation of Au-Al compounds by including H_2 in the device package. They postulated that the H_2-filled vacancies in the aluminum prevented or slowed the aluminum diffusion into the gold. Although interesting from a theoretical point of view, there is the possibility that hydrogen in the package will combine with any oxygen present (from ambient gas or on the cavity surface), to form water. Since water formation can result in corrosion, this procedure is not generally used. Also, packages often contain nickel, iron, and sometimes palladium (in Pd-Ag thick films). Any free hydrogen would soon be absorbed and embrittle these metals.

Aluminum bonds to thick-film gold metallization have always been more subject to failure by Kirkendall voiding than aluminum bonds to thin films, presumably because thick films contain more grain boundaries, vacancies,

TABLE I-2
FAILURE MODES ASSOCIATED WITH INTERMETALLIC FORMATION (CONDENSED FROM PHILOFSKY [I-2])

Symptom	Cause	Remedy
Open metallization around bonding pad (A) Zero pull strength—bond peeled off pad—fracture surface purple *(B)*	Voiding in $AuAl_2$	Keep circuit below 400 °C
Zero pull strength—bond peeled off pad—fracture surface tan (A) and *(B)*	Voiding in Au_5Al_2 Thermal cycling will aggravate	Make metallization thinner or reduce time at temperature
Zero pull strength—break at heel, fracture surface-tan (A) —tan or purple *(B)*	Intermetallic formation in heel of bond fatigues during thermal cycling	Make metallization thinner or use thicker wire or reduce time at temperature

and impurities. In the late 1970s, palladium was added to gold thick films for use in Al ultrasonic bonding [I-15,16]. This resulted either in a relatively stable Au-Al-Pd ternary compound or a concentration of Pd at the interface[3] that slowed both the Au and Al diffusion and lengthened the life of aluminum wire bonds. Many applications requiring aluminum-wire bonding currently employ such Pd-doped gold thick films. An attempt was made to utilize Pd-doped Al wire to improve the lifetime of bonds to plated and vacuum-evaporated gold thin films [I-16]. Unfortunately, the Pd affected the metallurgical characteristics of the wire leading to uncertain bonding results. Work along these lines with different Pd-Au alloys is continuing in several laboratories, particularly for use in ball-bond-bumped TAB.

2.2 IMPURITY ACCELERATED GOLD-ALUMINUM BOND FAILURES

The previous section described metallurgical diffusion, intermetallic compound formation, and Kirkendall voiding in pure bulk welds and wire bonds that can result in gold-aluminum weld failures. Horsting [I-17] was the first to discover that voiding-type wire-bond failures can be accelerated by impurities. He found that a number of impurities (e.g., Ni, Fe, Co, B) in gold-plated films may result in rapid intermetallic-appearing aluminum wire-bond failures. His model proposed that during a high-temperature bake, for a pure gold-aluminum bond, the intermetallic diffusion front moves through the gold plating down to the nickel under-plating, and the bond remains strong. For impure gold, the impurities became concentrated ahead of the intermetallic growth. At some concentration, precipitation of the impurity occurs. These particles then act as sinks for vacancies produced by the diffusion reaction, resulting in Kirkendall-like voids and leading to weak or zero-strength bonds. He introduced a thermal-stress test (390 °C for 1 hour followed by a pull test) as a pragmatic means of detecting gold films containing impurities. Horsting's failure model was derived from plated films and is treated more completely in the section on gold plating failures, see Figure P-1. Comparisons of his and other thermal-stress tests for bond reliability are given in the section on testing, Table T-1.

After Horsting's work showed that contaminants can accelerate bond failure, a number of other contaminants in gold-plated films, as well as from plastics, ambient atmospheres, and other sources, have all been shown to degrade bond reliability.

2.2.1 Effect of Halogens on the Gold-Aluminum Bond System

Halogens are pervasive and are well known to corrode aluminum metallization in integrated circuits [I-18]; see Appendix I-2. However, the first

[3]The literature is not clear on which occurs. More work is needed.

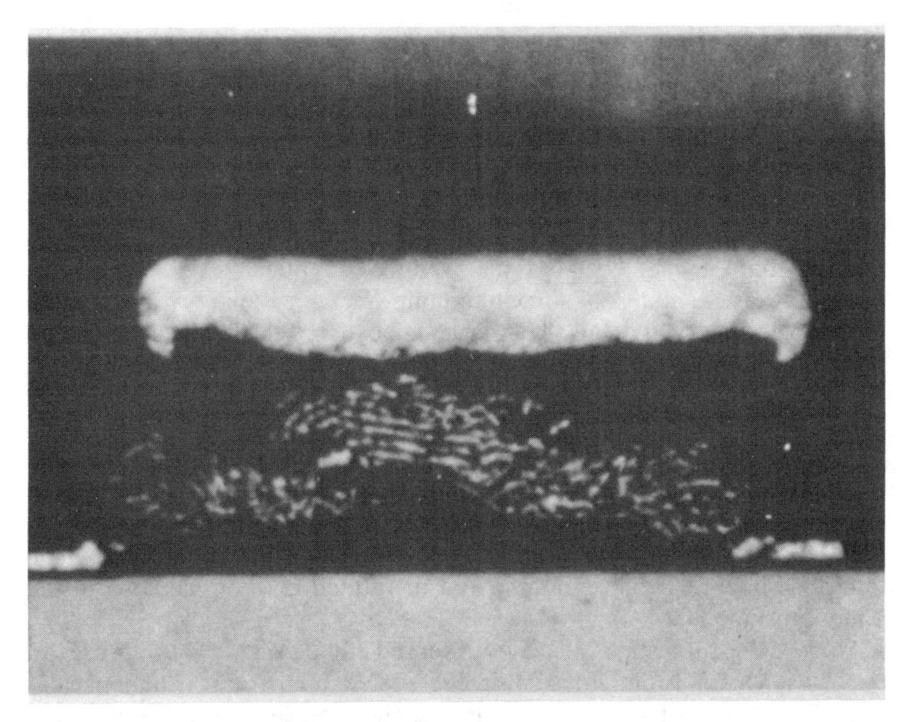

Figure I-10. A lamellar structure in the intermetallic region for 32-μm diameter gold thermocompression wire bonds to 1-μm thick aluminum metallization. This bond was aged at 200 °C for 24 hours in the presence of epoxy outgassed products. This structure is typical of two-phase regions commonly observed for eutectic and eutectoid microstructures (i.e., Pearlite) (after Thomas et al [I-19]).

observation that halogen compounds could degrade the strength of previously made gold bonds on aluminum metallization was by Thomas et al [I-19]. They cured various epoxies in the caps of TO-18 headers and sealed them to the package base which contained wire-bonded devices. Groups of these sealed packages were then stored at 150, 180, and 200 °C for up to 1,000 hours. Massive wire-bond failures occurred within 24 hours at 200 °C in devices with epoxies containing brominated flame retardants (tetrabromobisphenol-A). The Au-Al bonds failed after developing a weak lamellar microstructure as shown in Figure I-10. This structure is not characteristic of normal, intermetallic growth, but is more characteristic of a single-phase alloy that has grown unstable and separated. Apparently, the outgassing products from the epoxy attacked the intermetallic compound, diffusing in from the sides or other areas where the compound was exposed. No corrosion of the aluminum-bond pad material outside the bond area was observed. Controls with no epoxy in the caps resulted in strong bonds having normal, intermetallic growth. No failures of aluminum-wire bonds to the aluminum metallization in brominated epoxy-capped devices were observed.

Analysis of the outgassed products from these epoxies showed the presence of methyl bromide and ethyl chloride. Additional experiments verified that each of these gasses produced identical lamellar-structure bond failures. In addition, some aluminum-metallization corrosion by the gasses was observed. Thus, Thomas et al observed this lamellar-structure bond failure mechanism occurring with both Cl and Br containing gasses. Richie and Andrews [I-20] found the same structure for devices that had been exposed to CF_4/O_2 plasma treatment (100 watts, 1 Torr, 5-30 minutes). Their devices were die-bonded (both eutectic and epoxy), molded in plastic, and autoclaved [121 °C, 10,545 kg/m^2 (15 psi) steam]. Their experiment showed that fluorine will also produce the weak-lamellar structure at the gold-ball to aluminum-pad interface. It also revealed more rapid synergistic failures when Cl (from contaminated die-attach epoxy) was also in the bond interface.

Many other investigators have observed rapid Au-Al bond failures in the presence of brominated resins, elevated temperature, and usually humidity. Not all have reported finding the lamellar-intermetallic structure. Gale [I-21] found gross void formation and postulated that the voids may result from aluminum removal in the form of volatile halides. She found the activation energy for mechanical bond failure due to brominated epoxies to be 0.8 eV. Khan and Fatemi [I-22], however, found much lower activation energies for resistive bond failure (as opposed to mechanical) ranging from 0.2 to 0.5 eV. They applied the brominated resin directly to the bond areas. They also gave a chain of chemical reactions that can lead to the resistive bond failures. Their conclusions were that most of the reaction occurred with free bromine ions and that if the resins were purified of them, then failures, while not eliminated, would be dramatically reduced.

There is still a lack of agreement in the literature in both observations and interpretation {e.g., a corrosion mechanism producing $Al(OH)_3$ [I-22], metallurgical (phase separation) [I-19], oxidation of the aluminum in the intermetallic (Al_2O_3) [I-23], and volatile metal halide removal [I-21]}. It is quite possible that all of these mechanisms occur but under conditions that have not been clearly defined.

The role of H_2O in the bond-degradation process is not clear. Even if no autoclave is used, the high-temperature (approximately 180-200 °C) breakdown of epoxy encapsulants releases water [I-24]. Thus, the Thomas et al [I-19] sealed-device experiment presumably contained enough released

H_2O to affect the results. Their introduction of pure gasses however, presumably was dry. Perhaps H_2O serves as a catalyst or oxidant causing Au-Al voiding or lamellar structure to proceed at the lower temperatures of an autoclave. A summary of bond failures resulting from halogens is given in Table I-3.

2.2.2 Recommendations for Removing or Avoiding Halogen Contamination

It has been well established that halogens in an Au-Al bond interface or even in the environment after bonding can degrade Au-Al bond strength. All quoted experiments were run with halogen-free controls, and all controls survived much longer, at any temperature and humidity, than the contaminated devices. Halogens from wafer processing may become chemically bound to the aluminum, usually causing a brown appearance [I-25] which is difficult to remove. Pavio et al [I-26] were able to remove the fluorine coloration and restore bondability by Ar plasma sputtering (unspecified parameters). However, radiation damage (see Appendix C-1) destroyed their device's electrical characteristics. Most normal devices should survive such cleaning. Kern [I-27] used acetone (30-second rinse) to remove F^- ions from silicon surfaces. The effectiveness of this cleaning method has not been verified on bond pads and should be investigated. Chlorine has been removed from bond pads, at the wafer level, by heating the wafers to 300 °C in O_2 for 30 minutes [I-28]. This method is not generally applicable at the chip or packaging level. Outgassed halogens (mostly chlorine) from epoxy die attach are generally not chemically bound to the bond pads and ***are readily removable by plasma cleaning***.

There has been considerable study of fire-retardant molding compounds for the plastic encapsulation of devices. Such compounds degrade bonds, but some, with the same Br content, are far worse than others. This is presumably related to the free Br^- content of the resin. For most of the bond degradation to occur, high temperatures or pressures and humidities are required, and these are not a normal part of a device's life, so most devices will not be degraded. At any rate, there is little that can be done to prevent the bromine-induced bond degradation at the assembly level. The newest epoxy encapsulants have little free Br^- and appear to cause negligible bond degradation until the epoxy is degraded by high temperatures (approximately 250 °C). Thus, the problem is being solved.

2.2.3 Non-Halogen Epoxy Outgassing Induced Bond Failures

There have been reports of Au-Al wire-bond failures resulting from non-halogen epoxy die-attach outgassing products or other organic contami-

TABLE I-3

PROBLEMS FROM HALOGENS ON BOND PADS

	Source of Material	Contributing Causes	Negative Effect on Wire Bonds*	Corrective Action	Ref, I-
Wafer Processing	Silox etch (Fluoride)	Static DI wash	B,R	Agitated DI wash	25
	F or Cl residue on pads from RIE	May leave fluorocarbon polymer films, F or Cl	B,R>6 atomic % R<6 atomic %	Possible argon sputtering	26,28, 29
	Photo resist stripper	Dichlorobenzene residue	B,R,C	Complete removal	25
	Wafer sawing in city water	Cl in water	B,R	DI water w/surfacant	25
Cleaning	Trichlorethane (TCA)	Water contamination releases HCl	B,R,C	Use trichloroethylene (TCE) or better-plasma clean	25
	CF_4/O_2 plasma clean	Autoclave	R,C	Use O_2 or Ar plasma	20
Packaging	Cl from burn-in oven chloroprene gasket	Copper bonded gold thick film,surface $Cu \rightarrow CuCl_2$ (Al wire bonds)	R	Change gaskets to non-halogen elastomer	30
	Cl from plastic encapsulant	85°C / 85% RH, autoclave	R	Use plastic < 10 ppm Cl	31
	Br from encapsulation fire retardant	High temp (175-200°C) or 125°C autoclave	R	Avoid autoclave, high temp., or free Br.	19,21-24,32, 33

*B - Reduces bondability R - Reduces reliability C - Corrosion failures

nation [I-25,34 to 36]. Problems resulting from these products were very elusive, because failures occurred only occasionally, and made failure analysis as well as a full understanding very difficult. In one case [I-35], bond pads were directly exposed to the epoxy solvents and reactive diluants. It was found that the reactive dilutants caused organic deposits on the bond pad, which sometimes polymerized, reinforcing the oxide layer and preventing optimum bonding. The weak, poorly welded bond would then fail rapidly due to rapid vacancy diffusion through the microwelds or by corrosion (see Appendices I-1,2). Such problems can be detected initially by the use of a ball-shear test or prevented by plasma or UV-ozone cleaning before bonding (see section on cleaning).

2.3 INTERMETALLIC EFFECTS IN NON GOLD-ALUMINUM BONDS

2.3.1 Aluminum-Copper Wire Bond System

Recently, wires and metallizations other than those of Au or Al have been introduced. Copper-ball bonding to IC metallization has received considerable attention [I-37 to 42] for reasons of economy and also because of its resistance to wire sweep during plastic encapsulation. Copper is harder than gold, and thus more care is required during bonding to avoid cratering (see section on cratering and also Table C-2 to compare the metallurgical properties of Au and Cu). The hardness results in a tendency to push the softer aluminum aside [I-42] requiring harder metallization such as that described in reference [I-38]. Some bonding-machine parameters for Cu ball bonding were given in Figure C-3 (in the cratering section) and in each of the references [I-37 to 42].

The hardness of the copper also presents a problem in bonding to modern thin metallizations. Since copper oxidizes readily, ball bonds must be formed in an inert atmosphere, requiring modification of the bonder. Most studies of copper-ball bonding have been concerned with ball formation, bondability, and cratering as opposed to long-term reliability. The Al-Cu phase diagram shows the existence of five intermetallic compounds favoring the copper-rich side. Thus, there is the possibility of various intermetallic failures similar to those of the more familiar Al-Au system. Olsen and James [I-43] studied the thermal aging effects of Al wire bonded to OFHC[4] copper metallization. They found completely different aging characteristics, depending on the ambient. When the bonds were thermally aged in air, at 150 °C for 1,600 hours, they remained strong. However, aging in vacuum resulted in a rapid decrease in bond strength in the time

[4]Oxygen free high conductivity.

frame between 1,200 and 1,600 hours with an activation energy of 0.45 eV. They found that even though intermetallic compounds grew at the same rate as in vacuum, Cu oxide prevented or inhibited the growth of void-like grooves under the bond, increasing bond reliability.

Intermetallic growth has been studied by Atsumi et al [I-39] and Onuki et al [I-41]. Both found that the growth rate was less than half that of Al-Au bonds. The latter group found only $CuAl_2$ and CuAl compounds in bonds aged (apparently in air) at 150-200 °C. The activation energy for this growth was reported to be 1.2 eV. They reported no Kirkendall voiding but rather a weakening of the shear strength due to growth of the brittle $CuAl_2$.

Studies of Cu ball-bond strength in the presence of Br flame retardant and Cl in plastics were reported in [I-41]. The authors aged the premade bonds in proximity to epoxies similar to the studies of Thomas et al [I-19]. The Cu-Al bonds were strong after 1,245 hours at 200 °C, whereas Au-Al bonds failed after 700 hours.

The copper-aluminum bond system appears to be adequately reliable as long as some oxygen is present in the package. Presumably, enough oxygen is available in both nitrogen-filled hermetic packages [I-43] and plastic-encapsulated devices to result in a reliable bond system. The amount of oxygen should be limited so that oxidation of the wires [I-42] will not be a long-term reliability problem. Note that residual oxygen may be removed from hermetic packages if they have hydrogen containing gold electro-platings or are sealed with plated metal lids (see section on Plating). Eventually, free hydrogen will combine with residual oxygen, and the copper-aluminum bond will not be protected.

The bondability of the Cu-Al system is apt to be more of a problem than reliability, particularly with thin (6000-Å) metallizations where the hard Cu ball may push the metallization aside. Cratering (because of Cu-ball hardness) is also a potential problem. However, if Cu bonds do go into high production, it is possible that reliability degradation, due to some new unanticipated contaminant, could occur. For instance, sulfur compounds have long been known to corrode copper. Devices using Cu-Al bonds should be tested in such ambients (see Appendix IA-2). Aluminum metallization containing copper-aluminum intermetallics corrodes with Cl contamination and water (see section 2.3.6). In production, such corrosion may become a factor in copper-aluminum bonds even though Ounki et al [I-41] did not observe such problems in laboratory experiments.

2.3.2 Copper-Gold Wire Bond System

The current use of gold wires bonded to bare copper-lead frames and to Cu thick films in hybrids has led to interest in the reliability of this metallurgical system. The phase diagram shows three ductile intermetallic phases (Cu_3Au, AuCu, Au_3Cu) with overall activation energies of 0.8 to 1 eV. Kirkendall-like voiding has also been reported [I-44 to 46]. Temperature-time studies of thermocompression lead-frame bonds [I-47] in both air and vacuum show a significant decrease in strength as a result of void formation. Figure I-11 (based on a 40% bond-strength decrease) predicts a life of about five years at 100 °C. The lifetime would be longer if the failure criteria had been greater than a 40% decrease in strength. In either case, it appears to be adequate for most commercial devices. Pitt and Needes [I-48] studied gold TS bonds to thick-film copper and found little strength degradation at 150 °C for up to 3000 hours, and no failures at 250 °C over this period.

The reliability of Au bonds on Cu under autoclave (1000 hr), temperature cycling (-65 to 150 °C for 8000 cyc.), and temperature aging (150 °C, 1000 hr) was adequately verified for commercial devices by Lang and Pinamaneni [I-49]. The bond strength of Au-Cu bonds is apparently influenced by the microstructure, weld quality, and impurity content of the copper. Perhaps the greatest problem in bonding to copper-lead frames is assuring adequate cleaning (grease and copper oxide removal) [I-49,50] before and during bonding. In addition, if polymer die attach is used on lead frames, the polymer must be cured in an inert atmosphere to prevent oxidation while still maintaining significant gas flow to carry off the plastic outgas products. If this is not done, then both the chip and the lead-frame bondability or reliability will decrease as discussed in the previous section.

2.3.3 Silver-Aluminum Wire Bond System

Silver is used as a bond-pad plating on lead frames [I-51,52] and as metallization in commercial thick-film hybrids (usually in alloy form with Pt or Pd) [I-53]. Silver has also been tried as a substitute for gold wire for ball-bonding integrated circuits [I-54].

The Ag-Al phase diagram is complex with numerous-intermetallic phases. However, only the intermediate mu and zeta phases have been observed in Ag-Al wire-bond interfaces [I-52,55]. The intermetallic phases had an activation energy for growth of 0.75 eV. Kirkendall voids have been observed in this metal system [I-56,57], but generally at higher temperatures than experienced by microelectronic circuits. Failure of Ag-Al electrical contacts due to interdiffusion was first pointed out by Hermansky [I-57]. Jellison [I-55] was the first to observe the rapid humidity corrosion degradation mechanism in wire bonds, and this has become the major reason why the Ag-Al metal combination is seldom used. Chlorine was

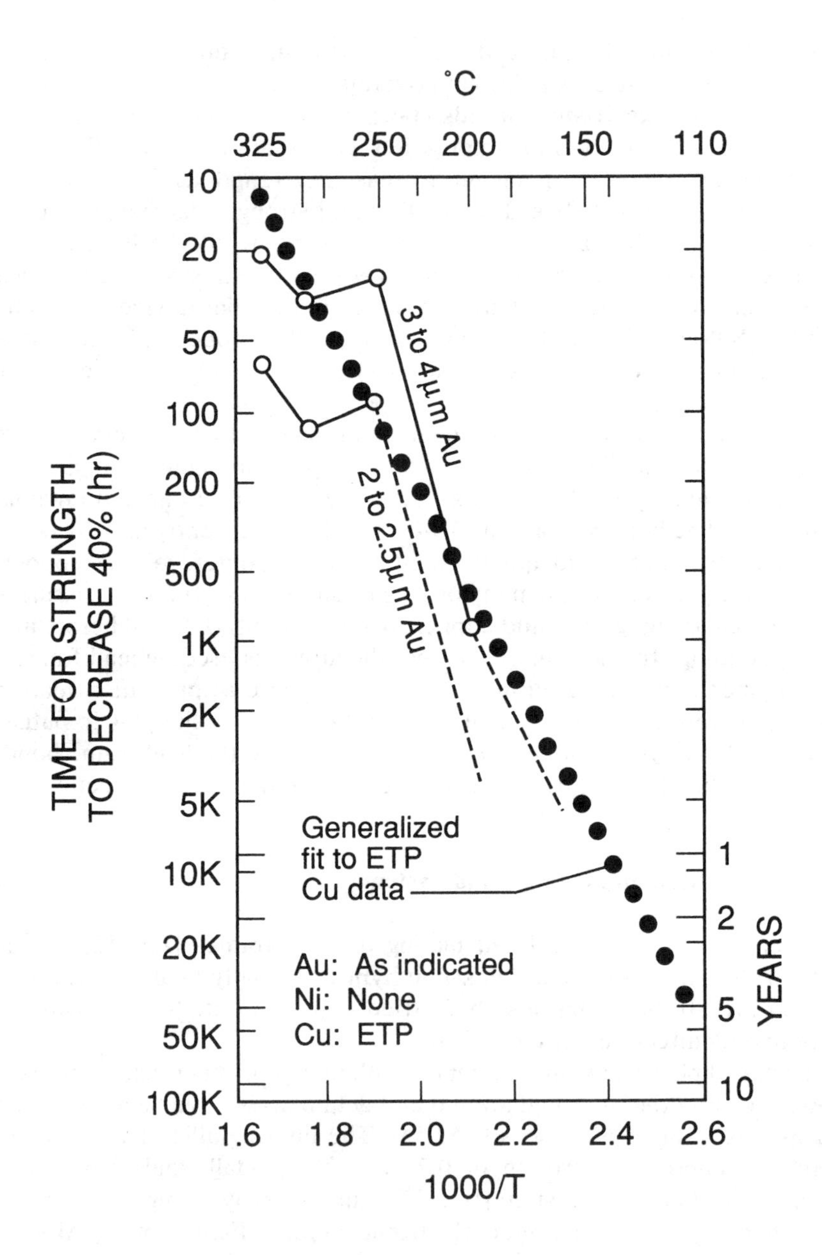

Figure I-11. Temperature dependence of time to decrease copper-gold bond strength 40% below the as-bonded strength. Dotted line represents generalized fit to the Cu data (after Hall et al [I-47]).

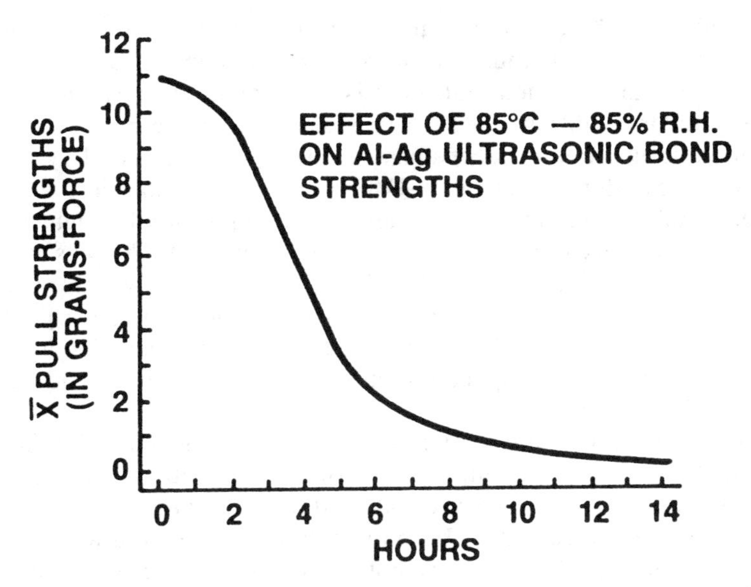

Figure I-12. Degradation of aluminum-wire bonds to silver-plated surfaces by temperature and humidity. Each data point represents $\bar{X}$ of 10 wire pull-test strengths for 99.99% Al wire of 50-μm diameter (after James [I-51]).

identified as the main driving element of the corrosion process.[5] When it was removed from the silver by an NH_4OH-H_2O ultrasonic-cleaning bath, subsequently made Al-Ag bonds remained reliable. The humidity-corrosion mechanism has been extensively verified [I-51,54,58]. Wire-bond strength degradation of Al-Ag wire bonds was studied by James and is shown in Figure I-12. Failure analysis revealed $Al(OH)_3$ in the failed bond interface. This led to the proposal that failure occurred by the classical aluminum-chlorine corrosion mechanism which regenerates the chlorine to continue the reaction (see Appendix IA-2).

In other work, sections of bonds revealed that the corrosion actually took place in the zeta intermetallic phase [I-54]. It is not clear why the silver-aluminum bond system corrodes more readily than the Au-Al system, unless the silver or its oxide acts as a catalyst or an intermediate reaction exists in addition to the galvanic reaction.[6] The activation energy

[5]Presumably other halogens would cause the same effect.

[6]In the electrochemical series, Au^+ is more positive than Ag^+, and one would assume the Au-Al couple would corrode as readily as the Ag-Al couple. However, Ag^{++} is more positive than gold. No similar information is available on the conductive oxides of silver.

for bond strength degradation due to this corrosion process was reported to be 0.3 eV [I-58] and thus, it is not strongly temperature dependent. Two authors [I-51,55] verified that there is no comparable corrosion reaction under similar conditions for Al-Au bonds or for Au-Ag bonds.

An additional failure mechanism of the Ag-Al wire-bond system was described by Shukla and Singh-Deo [I-58]. Aluminum wires bonded to silver metallization in CERDIPs failed catastrophically by high electrical resistance (not because they were weak mechanically). This was attributed to a selective oxidation of the Ag-Al intermetallic layer which resulted in an insulating-oxide barrier in the interface. The activation energy for this reaction was given as 1.4 eV, and the mechanism was active above 400 °C. This would not be expected to affect devices processed and operated at normal temperatures.

Large-diameter aluminum wires are routinely bonded to Pd-Ag thick-film metallization in automotive hybrids [I-53]. However, preparation requires washing with solvents, followed by careful resistivity-monitored cleaning in deionized water. Then the hybrids are covered with a silicone gel for further protection. It is not clear whether the Pd additive, the careful cleaning [I-51,55], the silicone gel, or a combination completely prevents the Al-Ag bond-corrosion problems described above. Each is helpful, but until the corrosion mechanism is fully understood, the use of thick-film silver for Al bonding should be undertaken with caution, qualifying the devices in 85 °C/85% RH.

2.3.4 Aluminum-Nickel Wire Bond System

When the price of gold (platings) increased dramatically during the 1970s, nickel coatings were substituted for gold on power devices. Large-diameter aluminum wires were easily bonded to the nickel and were found to be reliable under the various environments. These in-house studies by device manufacturers apparently were not published in the open literature, so most information has been obtained informally.

Large-diameter, >75-μm (3-mil), aluminum wires bonded to nickel platings or inlays have been used in high production on power devices for over fifteen years with no significant reliability problems reported. In most cases, the nickel is deposited from electroless boride or sulfamate solutions. Low-stress films electroplated from sulfamate baths also result in reliable bonds. However, phosphide electroless nickel solutions that co-deposit more than 6 or 8% of phosphorus can result in both reliability and bondability problems.

Aluminum-nickel bonds are more reliable than Al-Ag or Al-Au bonds. The Al-Ni phase diagram is complex with numerous intermetallic phases and transitions. The system is generally refractory and is used in high-temperature applications such as aircraft turbine blades. Apparently the activation energy for the growth of these phases is high (>1 eV, from melting point data), and Kirkendall voiding does not take place at temperatures and times encountered by power devices. The electrochemical series indicates that most Ni^+ reactions have a negative potential (Al^+ has a negative potential also), whereas both gold and silver are strongly positive. Thus, galvanic corrosion is less likely with Ni-Al bonds.

The main difficulty encountered when bonding to nickel platings is bondability rather than reliability. Nickel surfaces oxidize, producing the same bondability problems discussed in section 3.7 on plating. Thus, packages should be bonded soon after they are Ni-plated, protected in an inert atmosphere, or chemically cleaned before bonding. Changing bonding machine schedules, such as impacting the tool-wire-plating with the ultrasonic energy applied, has been reported to improve bondability to slightly oxidized Ni surfaces. Various surface preparation techniques (such as sandblasting) are sometimes applied before or after Ni plating to increase bondability.

2.3.5 Gold-Gold, Aluminum-Aluminum, and Gold-Silver Wire Bond Systems

The gold-gold wire-bond system is extremely reliable. It is not subject to interface corrosion, intermetallic formation, or other bond-degrading conditions. A poorly welded Au-Au bond will improve in strength with time and temperature [CL-10]. Although cold ultrasonic Au-Au wire bonds can be made (using cross-grooved or special-surfaced bonding tools), gold welds best with heat. Either thermocompression or thermosonic bonds are easily and reliably made. Thermocompression bondability, however, is strongly affected by surface contamination (see section on cleaning and reference [CL-10]).

The aluminum-aluminum wire bond system is also extremely reliable. It is not subject to interface corrosion any more than the aluminum bond pad is, and in a corrosive environment the surrounding pad will often be digested, but the bond will be intact. While thermocompression Al-Al bonds can be made with high deformation, aluminum welds best ultrasonically, and heat does not significantly improve the weld quality. Studies similar to Jellison's [CL-10] for Au-Au bonding have not been made to determine that Al-Al bond interfaces get stronger with temperature and time. However, such interfaces do not weaken, as evidenced by the reliability of large numbers of Al-Al bonds in CERDIPs and other devices that are exposed to high temperatures.

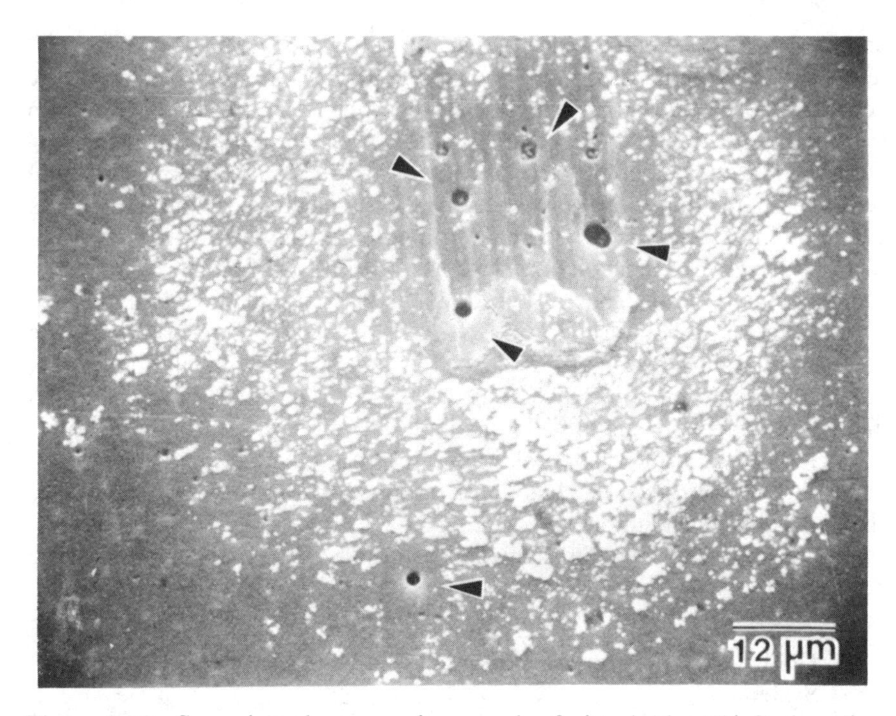

Figure I-13. Scanning electron micrograph of pitted Al-1.5% Cu bonding pads following ball shearing showing the lack of bonding (i.e., no gold residue or any evidence of bonding) in the corrosion halo areas as compared with the normal areas (after Thomas et al [I-60]).

The Au-Ag wire-bond system has been shown to be reliable over very long times at high temperatures by James [I-51]. No interface corrosion has been reported and no intermetallic compounds form. Gold-wire bonds to silver-plated lead frames have been successfully used in high production for many years. Bondability problems can result if the silver plating is heavily tarnished by sulfur compounds, but this tarnish can be prevented or removed. To increase the bondability of silver plating in high production, TS bonding is often performed with stage temperatures in excess of 250 °C which dissociates thin silver-sulfide films.

2.3.6 Aluminum Metallization Containing Copper

Aluminum integrated-circuit metallization often contains several percent copper to inhibit electromigration. Isolated copper-aluminum intermetallic aggregates can form if the sintering (heat treatment) is incorrect. These aggregates have different electrochemical potentials from pure Al,

TABLE I-4
WIRE BOND INTERFACE RELIABILITY

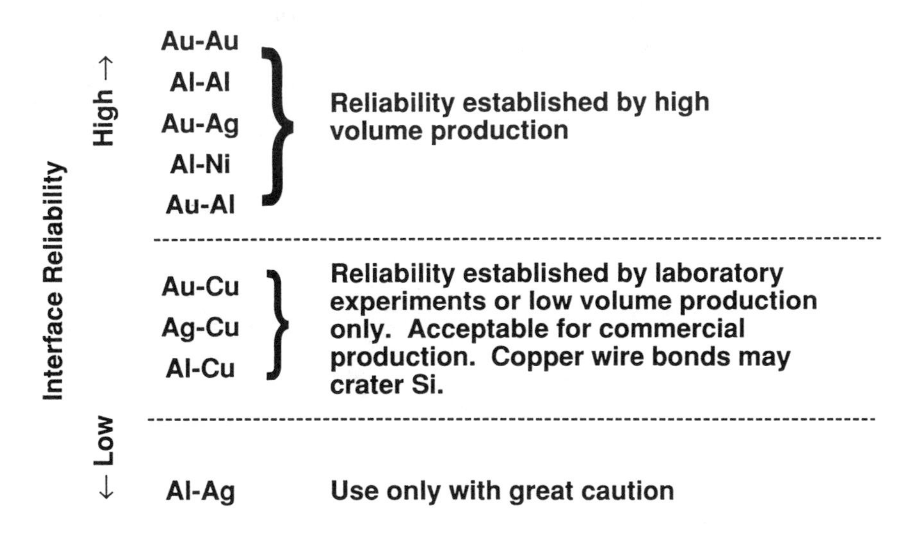

and also from other Cu-Al phases ($\Delta \approx 0.1$ V) [I-59]. Corrosion is usually attributed to the theta phase (Al_2Cu) aggregates. The combination of moisture and traces of a halogen (that is usually present) will result in corrosion on the bond pad, and will often discolor the metal (see Appendix IA-2). This condition is sometimes referred to as the ''brown'' or ''black'' metal problem [I-25,60].

The corrosion can result in both bondability and reliability problems when bonded with gold wire. Figure I-13 is an example of a pitted-bond pad that resulted in weak TS bonding. The problem can be eliminated during processing if a homogeneous distribution of copper is obtained or by careful cleaning to remove halogens after such manufacturing steps as wafer sawing and washing. If ''brown'' metal is observed during the assembly operation, such chips should not be used.

The addition of copper to aluminum bond pads makes them harder to bond, requiring more ultrasonic energy during bonding. This can increase the probability of cratering. Cratering may be reduced by new high-impact bonding schedules (see the cratering section). If copper is diffused to the bond-pad surface, it will oxidize and reduce bondability. In general, some loss of bondability is experienced when the copper content of aluminum metallization is increased over approximately 1.5%. The copper content, as well as the proper heat treatment of the IC metallization to minimize its effects, cannot be controlled at the assembly and packaging level.

However, it is necessary for packaging personnel to understand the potential problems when dealing with chips containing Cu in the metallization.

Table I-4 shows the general relative reliability of the various metallurgical bond systems discussed above.

2.4 APPENDIX IA-1

RAPID BOND FAILURE IN POORLY WELDED GOLD-ALUMINUM WIRE BONDS

Bond failures, due to Au-Al intermetallic growth and void formation, were described earlier in this section. Such descriptions are usually based on the concepts of bulk Au-Al couples, often with the synergistic effect of halogens and moisture. However, impurity-free but poorly welded bonds have been observed to fail much more rapidly than strongly welded bonds [I-16,35,54,64].

In order to explain this, it is necessary to understand early (immature) bond formation. The initial welding between surfaces consists of a series of isolated microwelds, examples of which are shown in Figure IA-1 for early Al-Al ultrasonic bonds [I-65] and Figure IA-2 for Au-Au TC bonds. Such initial welding for US and TC bonds takes place near the perimeter where the metal motion (deformation) is maximum. This deformation sweeps surface oxide and contaminants aside into debris zones, allowing intimate contact between the two metallic parts.

The exact ultrasonic welding mechanism is not fully established, but presumably the same energy transfer mechanism that softens the metals,

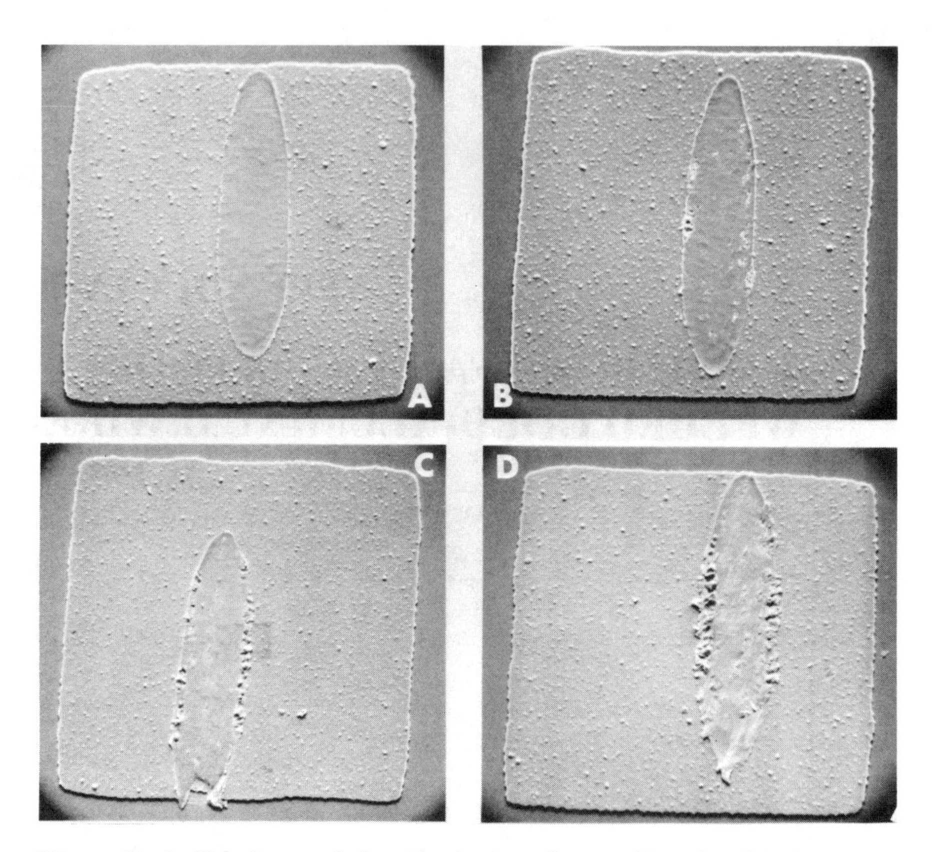

Figure IA-1. Bond growth for Al wire bonding to Al pads with the aluminum wire lifted off (lift-off patterns). (A) Zero weld time (no ultrasonic energy), (B) 4-ms weld time, (C) 7-ms weld time, and (D) 10-ms weld time (after Harman and Leedy [I-64]).

without significant heat (about 50 °C temperature rise), also supplies the required activation energy to form the metal-to-metal chemical bonds. As the weld matures, the microwelds grow, join together, and spread inward, sometimes leaving the center of the weld unbonded [I-65]. Gold-ball thermosonic bonds are somewhat different and have been observed to either start in the center and spread outward or to be primarily welded near the perimeter.

Contaminants, such as noncorrosive organic films [I-35] and unremoved glass on the bonding pad [I-64], can limit the growth and spreading of the initial microwelds. Thus, immature bonds, consisting of isolated microwelds, can result either from poor bonding machine setup (underbonding),

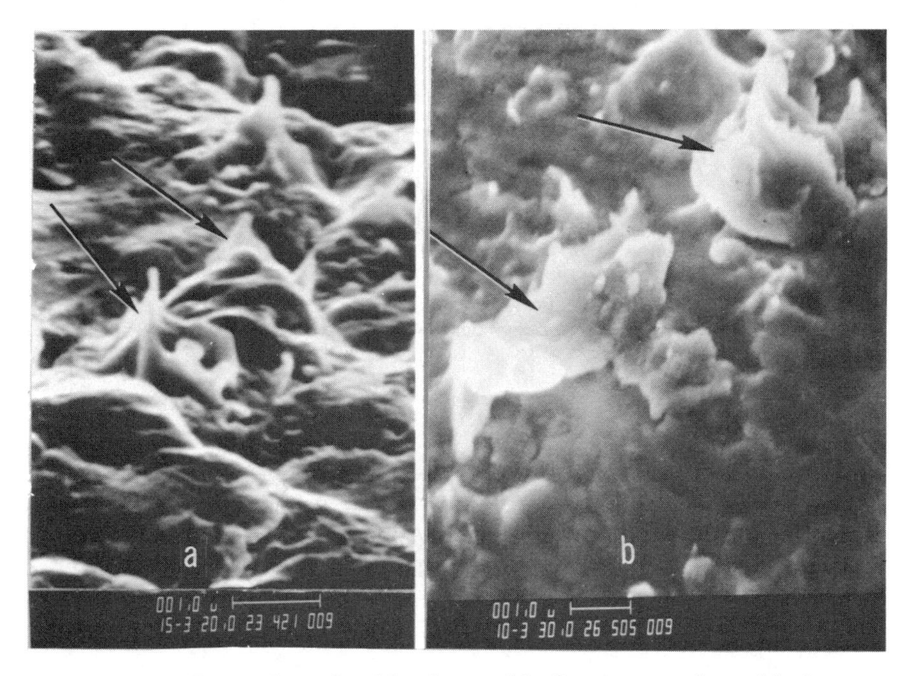

Figure IA-2. Examples of gold microwelds from a poorly welded, separated, bond interface (after Harman [I-67]).

or from some form of contaminant in the interface that prevents intimate contact between surfaces, except in areas of highest deformation.[1] A two-dimensional finite element model for Au-Al diffusion in structures with microweld dimensions was developed by Wilson [I-66].

This model gives an explanation for the observed rapid bond failures in poorly welded Au-Al structures. Metallic diffusion takes place by way of defects. Surfaces and grain boundaries result in rapid diffusion for several reasons, one of which is that they contain large numbers of defects. The actual stress due to bonding and excess defect level within these microwelds is unknown so assumptions had to be made. Microwelds are generally too small to contain several grains, and thus, internal grain boundaries. They are assumed to be one grain for modeling purposes. Thus, the model assumed that most vacancies were on the microweld surfaces as shown in Figure IA-3. The welded interface is assumed to contain the vacancies that were on the original surfaces. The diffusion coefficients between the

[1]Isolated microwelds offer more surface area for attack and are therefore more subject to halogen corrosion than strong uniformly welded bonds.

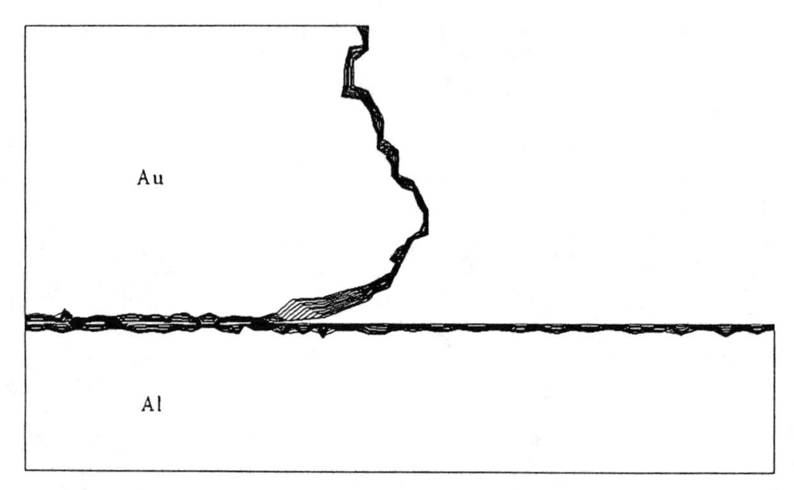

Figure IA-3. The vacancy distribution at t = 0 for Figure IA-6. This is typical for the other microweld shapes. Apparent variations of the vacancy densities are an artifact of the mesh.

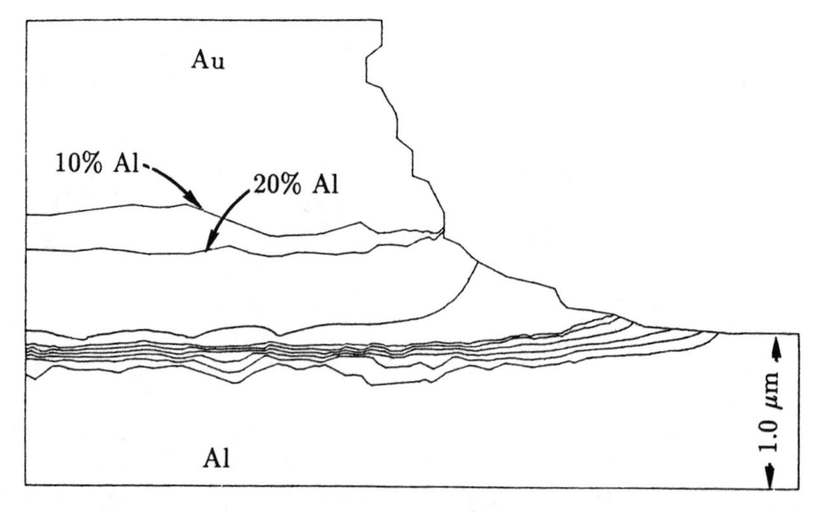

Figure IA-4. Intermetallic phases after diffusion for a smoothly graded weld. Each contour line represents 10% increase of the opposite metal diffusing to that position.

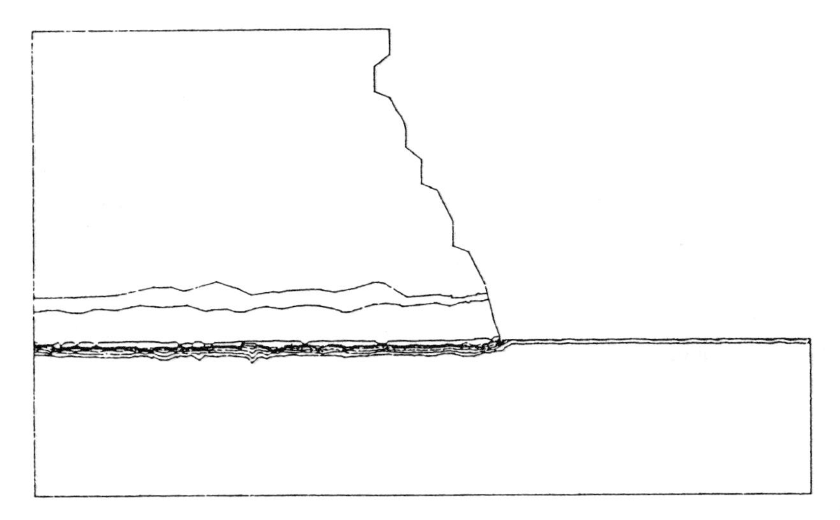

Figure IA-5. Intermetallic phases after diffusion for a relatively vertical weld.

Au, Al, and the five Au-Al compounds for layer growth were obtained from Philofsky [I-2,3]. Microweld shapes that were observed in actual failed bonds (Figure IA-2) were modeled as well as simple butt junctions. Results from equivalent time-temperature soaks of about 20 minutes at

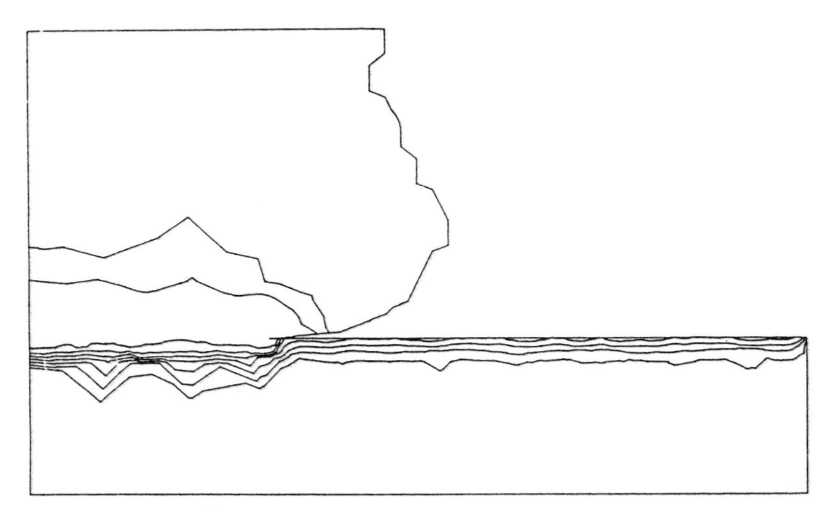

Figure IA-6. Intermetallic phases after diffusion for a sharply intersecting weld.

200 °C are given in Figures IA-4 through IA-6.

The first conclusion is that both the rate and nature of the diffusion is dependent on the shape of the microwelds. The almost vertical couple in Figure IA-5 most closely resembles bulk diffusion. In Figure IA-6, the sharp intersection of the Au and Al serves as a hot spot for vacancy diffusion. This in turn leads to rapid Au diffusion along the surface of the Al. In both Figures IA-4 and IA-6, diffusion into the other material is considerably more rapid than in Figure IA-5 (which resembles bulk couple diffusion). Thus, according to this model, diffusion can be several times more rapid in most microweld geometries than it would be in completely welded couples, supporting observations that poorly welded Au-Al bonds fail more rapidly.

2.5 APPENDIX IA-2

VARIOUS BOND-RELATED CORROSION REACTIONS

Halogen-Aluminum Corrosion Reactions

The general chlorine corrosion equations and explanations for aluminum metallization are given below from Paulson and Lorigan [I-18] and Iannuzzi [I-61]. (Similar equations have been given for chlorine [I-31] and for bromine [I-22].)

The corrosion mechanism consists of the adsorption of Cl^- on the oxide-solution interface under the influence of an electric field (caused by the electric double layer at the oxide-solution interface and/or the galvanic couple of the bond) in competition with OH^- or H_2O molecules for surface sites on the hydrated oxide surface. This is followed by the formation of a basic hydroxychloride aluminum salt with the aluminum oxide cations on the hydrated oxide surface:

$$Al(OH)_3 + Cl^- \longrightarrow Al(OH)_2\,Cl + OH^-. \qquad \text{(IA} - 1)$$

Once the surface oxide is dissolved, the underlying Al reacts with the Cl^- by the equation

$$Al + 4Cl^- \longrightarrow Al(Cl)_4^- + 3\,e^-. \qquad \text{(IA} - 2\text{)}$$

The $Al(Cl)_4$ will then react with the available water by the reaction

$$2AlCl_4^- + 6\,H_2O \longrightarrow 2Al(OH)_3 + 6H^+\ 8Cl^-. \qquad \text{(IA} - 3\text{)}$$

This process liberates the Cl^- ion, which is then available to continue the corrosion process via equations (IA-1) and (IA-2). In addition, the gold ball bond on aluminum produces a galvanic couple which can accelerate corrosion by acting as the driving force for the aluminum oxidation reaction. The region near the bond is an area of higher chloride ion concentration (compared to the overall surface) because of the reduction reaction of Cl_2 by the gold electrode. Although bond-pad corrosion (outside the bond) is not generally considered as a bond failure, the electrical resistance does increase, and the device becomes nonfunctional in a manner similar to that caused by Kirkendall voiding. However, many of the actual bond failures due to halogens ***are*** attributed to a corrosion mechanism similar to the above, but occurring to the aluminum under the bond, within the Au-Al intermetallic. The bond pad, away from the bond, is often corrosion-free because the concentration of free Cl^- is highest at the bond, due to the galvanic reaction.

One possible reaction for high-temperature brominated epoxy degradation of bond strength was proposed by Khan et al [I-22] with the reaction products of equation (IA-3). A different mechanism was proposed by Ritz et al [I-23] and is given below in equations (IA-4 to IA-7). The possible reaction (unbalanced) might include the liberation of Br from CH_3Br or HBr from the high-temperature breakdown products of the resin.

$$HBrCH_3Br \text{------}\!\!\longrightarrow CH_3^+ + Br^- \qquad \text{(IA} - 4\text{)}$$

$$\text{and } 4HBr + 2O \text{------}\!\!\longrightarrow 4Br^- + 2H_2O. \qquad \text{(IA} - 5\text{)}$$

Br^- would react with the Al in Au_4Al forming $AlBr_3$ and Au. The Al once extracted in the corrosion cell as $AlBr_3$, is easily oxidized. This oxidation reaction becomes the driving force until the Au_4Al intermetallic phase is consumed as in the equations below

$$Au_4Al + 3Br^- \text{------}\!\!\longrightarrow 4Au + AlBr_3 \qquad \text{(IA} - 6\text{)}$$

$$2\,AlBr_3 + 3O \text{------}\!\!\longrightarrow Al_2O_3 + 6Br^-. \qquad \text{(IA} - 7\text{)}$$

Note that equation (IA-7) produces aluminum oxide, whereas the normal halogen corrosion, equation (IA-3), produces the hydroxide. With the reaction being autocatalytic, the Br^- is freed to start the corrosion over again. These proposed reactions come closest to explaining the lamellar structure found in [I-20]. The equations are based on Au_4Al, which has a low occurrence, except as the ***final*** reaction product in a gold-rich couple (see Figure I-2 and I-3). Thus, the probability of this taking place may be low.

The most important result of all of these reactions is that ionized halogen is liberated at the end. Thus, only a small amount of the halogen is required to completely corrode a bond pad or a bond interface.

Sulfur-Copper-Chlorine Corrosion Reactions

Copper (as used for ball bonding and lead frames) can be corroded readily (or tarnished) by sulfur and sulfur compounds. Some chemical reactions, including the increased synergistic corrosion by Cl_2 and NO_2, are given in Table IA-1 (Abbott [I-62]). These are simple atmosphere reactions, not driven by galvanic couples such as may occur at a Cu-Al bond. Memis [I-63] has discussed copper corrosion by sulfur and that it led to failure in electronic packages. Sulfur and its gaseous compounds easily diffuse through silicone materials but, fortunately, some epoxy seals serve as an effective barrier and can prevent it from entering the package. Nevertheless, Al-Cu bond studies should include the atmospheres from Table IA-1 to establish the importance (or lack of it) of shielding the bonds from these omnipresent chemicals.

TABLE IA-1
FILM CHEMISTRIES FOR TARNISH FILMS ON COPPER [I-62][a]

Environment	Approximate Compositions in Percent		
	Cu_2O	CuO	Cu_2S
$H_2S - O_2 - H_2O$	7–10	4–5	80–85
$H_2S - SO_2 - O_2 - H_2O$[b]	40–50	10–15	30–35
$H_2S - NO_2 - O_2 - H_2O$	15–20	7–10	70–75
$H_2S - Cl_2 - O_2 - H_2O$	40–50	35–40	15–20
$H_2S - SO_2 - NO_2 - O_2 - H_2O$	55–60	30–35	10–20
$H_2S - SO_2 - NO_2 - Cl_2 - O_2 - H_2O$	50–55	30–40	15–20

[a]Reference [I-63] points out that elemental sulfur vapor has a high occurrence and will also corrode copper.
[b]200 ppb SO_2.

2.6 APPENDIX IA-3

ASSESSING THE POSSIBLE RELIABILITY OF PROPOSED NEW BOND METAL SYSTEMS

1. Can it be welded by ultrasonic, TC, and/or TS methods? As a first cut, see tables of ultrasonically bondable materials in ***The Welding Handbook***.[1] For device metallization, consider how any possible dopants (e.g., Cu and Si in Al) might affect bondability and cratering.
2. Are there any potential bondability or handling problems for high-volume production? Does the metallization form a soft oxide during long storage or normal chemical treatment. As an example, nickel and copper are bondable by aluminum and gold, but each has a soft oxide that reduces the bondability and must be removed or special bond schedule techniques developed. Aluminum is soft but has a hard brittle oxide which easily shatters and which is pushed aside into debris zones during bonding. This oxide does not present a bonding problem.

[1]The Welding Handbook, The American Welding Society, 7th edition, Vol. 3, is useful. The 8th edition will have a greatly improved section on ultrasonic welding and should be published in 1990.

3. Is the new wire harder than Al or gold? If so, the cratering potential is higher, and will require special bonding techniques or schedules (see section 5.1.5).
4. Are there numerous intermetallic compounds that may form and affect reliability? Look up the phase diagram. If such intermetallic compounds exist and have high melting points (e.g., >1000 °C), then they should not significantly affect reliability. If low (≲500 °C), then their reliability is in question. The activation energy of individual compounds crudely, can be obtained from:

$$\text{activation energy} \approx \frac{\text{melting point in °K}}{35 \text{ to } 40}$$

in K calories/mole (for face-centered cubic compounds), where 1 eV ≈ 23 kilocal/mole. There are many complications in using such a formula because the nucleation and other properties of different compounds may vary, but it is a beginning. As an example, nickel-aluminum compounds have a very high melting point. They are refractory and are stable in jet turban blades; Ni-Al bonds do not fail from intermetallic problems.
5. Are the individual materials easily corroded? Does the bond couple have a high probability of making a corrosion couple? Halogens and sulfur compounds are omnipresent, so look up their effect on the bare metals. Look up the electrochemical series potentials.[2] If the most common reduction reactions of the metals are widely separated, then corrosion is probable. As an example, Al is strongly negative and Au and Ag are strongly positive, thus there is corrosion on such bonds in the presence of halogens and moisture. Such corrosion has been discussed in this chapter of the book. On the other hand, most Ni reactions are negative except one, whose occurrence is less probable. Thus, the Ni-Al couple has not been observed to corrode. The electrochemical series must be used with caution since the measurements are made under very specific conditions. However, such information is indicative that this is true.

[2]Any chemical handbook or the Handbook of Chemistry and Physics (CRC Press) contains such tables.

2.7 REFERENCES

I-1. Philofsky, E., Intermetallic Formation in Gold-Aluminum Systems, Solid State Electronics 13, pp. 1391-1399 (1970).

I-2. Philofsky, E., Design Limits When Using Gold-Aluminum Bonds, 9th Annual Proceedings IEEE Reliability Physics Symposium, Las Vegas, Nevada, pp. 11-16 (1971).

I-3. Philofsky, E., Purple Plague Revisited, 8th Annual Proceedings IEEE Reliability Physics Symposium, Las Vegas, Nevada, pp. 177-185 (1970).

I-4. Hansen, M., The Constitution of Binary Phase Diagrams, Second Edition, McGraw Hill, New York, New York (1958).

I-5. Gerling, W., Electrical and Physical Characterization of Gold-Ball Bonds on Aluminum Layers, 34th Proc. IEEE Electronic Components Conference, New Orleans, Louisiana, May 14-16, 1984, pp. 13-20.

I-6. Weaver, C., and Brown, L. C., Diffusion in Evaporated Films of Gold-Aluminum, Phil. Mag. 7, pp. 1-16, 1961.

I-7. Schnable, G. L., and Keen, R. S., RADC Contract No. AF 30, 1966.

I-8. Kashiwabara, M., and Hattori, S., Formation of Al-Au Intermetallic Compounds and Resistance Increase for Ultrasonic Al Wire Bonding, Review of the Electrical Communication Laboratory 17, pp. 1001-1013, 1969.

I-9. Onishi, M., and Fukumoto, K., Diffusion Formation of Intermetallic Compounds in Au-Al Couples by Use of Evaporated Al Films, Jap. J. Metallurgical Soc., p. 38, 1974.

I-10. Chen, G. K. C., On the Physics of Purple-Plague Formation, and the Observation of Purple Plague in Ultrasonically-Joined Gold-Aluminum Bonds, IEEE Trans. on Parts, Materials, and Packaging PMP-3, p. 149, 1967.

I-11. Anderson, J. H., and Cox, W. P., Failure Modes in Gold-Aluminum Thermocompression Bonds, IEEE Trans. Reliability R-18, p. 206, 1969.

I-12. Charles, Jr., H. K., and Clatterbaugh, G. V., Ball Bond Shearing - A Complement to the Wire Bond Pull Test, Intl. J. Hybrid Microelectronics 6, pp. 171-186 (1983).

I-13. Majni, G., and Ottaviani, G., AuAl Compound Formation by Thin Film Interactions, J. Crystal Growth 47, pp. 583-588 (1979).

I-14. Shih, D-Y. and Ficalora, P. J., The Reduction of Au-Al Intermetallic Formation and Electromigration in Hydrogen Environments, 16th Annual Proc. IEEE Reliability Physics Symposium, pp. 268-272, San Diego, California. (1978).

I-15. Horowitz, S. J., Felton, J. J., Gerry, D. J., Larry, J. R., and Rosenberg,R. M., Recent Developments in Gold Conductor Bonding Performance and Failure Mechanisms, Solid State Technology 22, pp. 37-44, March (1979).

I-16. Hund, T. D., and Plunkett, P. V., Improving Thermosonic Gold Ball Bond Reliability, IEEE Trans. Components, Hybrids, and Manufacturing Technology CHMT-8, pp. 446-456 (1985).

I-17. Horsting, C. W., Purple Plague and Gold Purity, 10th Annual Proc. IEEE Reliability Physics Symposium, Las Vegas, Nevada, pp.155-158 (1972).

I-18. Paulson, W. M., and Lorigan, R. P., The Effect of Impurities on the Corrosion of Aluminum Metallization, 14th Annual Proc. IEEE Reliability Physics Symposium, Las Vegas, Nevada, April 20-22, 1976, pp. 42-47.

I-19. Thomas, R. E., Winchell, V., James, K., and Scharr, T., Plastic Outgassing Induced Wire Bond Failure, 27th Proc. IEEE Electronics Components Conference, Arlington, Virginia, May 16-18, 1977, pp. 182-187.

I-20. Richie, R. J., and Andrews, D. M., CF_4/O_2 Plasma Accelerated Aluminum Metallization Corrosion in Plastic Encapsulated ICs in the Presence of Contaminated Die Attach Epoxies, 19th Annual Proc. IEEE Reliability Physics Symposium, Orlando, Florida, April 7-9, 1981, pp. 88-92.

I-21. Gale, R. J., Epoxy Degradation Induced Au-Al Intermetallic Void Formation in Plastic Encapsulated MOS Memories, 22nd Annual Proc. IEEE Reliability Physics Symposium, Las Vegas, Nevada, April 3-5, 1984, pp. 37-47.

I-22. Khan, M. M., and Fatemi, H., Gold-Aluminum Bond Failure Induced by Halogenated Additives in Epoxy Molding Compounds, Proc. 1986 Intl. Symposium on Microelectronics (ISHM), Atlanta, Georgia, October 6-8, 1986, pp. 420-427.

I-23. Ritz, K. N., Stacy, W. T., and Broadbent, E. K., The Microstructure of Ball Bond Corrosion Failures, 25th Annual Proc. IEEE Reliability Physics Symposium, April 1987, San Diego, California, pp. 28-33.

I-24. Lum, R. M., and Feinstein, L. G., Investigation of the Molecular Processes Controlling Corrosion Failure Mechanisms in Plastic Encapsulated Semiconductor Devices, 30th Proc. IEEE Electronics Components Conference, San Francisco, California, April 28-30, 1980, pp. 113-120.

I-25. Nesheim, J. K., The Effects of Ionic and Organic Contamination on Wirebond Reliability, Proc. 1984 Intl. Symposium on Microelectronics (ISHM), Dallas, Texas, September 17-19, 1984, pp. 70-78.

I-26. Pavio, J., Jung, R., Doering, C., Roebuck, R., and Franzone, M., Working Around the Fluorine Factor in Wire Bond Reliability, Ibid., pp. 428-432.

I-27. Kern, W., Radiochemical Study of Semiconductor Surface Contamination, I. Adsorption of Reagent Components, RCA Review, June 1970, p. 224.

I-28. Lee, W-Y., Eldridge, J. M., and Schwartz, G. C., Reactive Ion Etching Induced Corrosion of Al and Al-Cu Films, J. Appl. Phys. 52 (4), April 1981, pp. 2994-2999.

I-29. Graves, J. F., and Gurany, W., Reliability Effects of Fluorine Contamination of Aluminum Bonding Pads on Semiconductor Chips, 32nd Proc. IEEE Electronics Components Conference, San Diego, California, May 10-12, 1982, pp. 266-267.

I-30. Forrest, N. H., Reliability Aspects of Minute Amounts of Chlorine on Wire Bonds Exposed to Pre-Seal Burn-In, Intl. J. Hybrid Microelectronics 5, pp. 549-551, November 1982.

I-31. Gustafsson, K., and Lindborg, U., Chlorine Content in and Life of Plastic Encapsulated Micro-Circuits, 37th Proc. IEEE Electronics Components Conference, Boston, Massachusetts, May 11-13, 1987, pp. 491-499.

I-32. Blish II, R. C., and Parobek, L., Wire Bond Integrity Test Chip, 21st Annual Proc. IEEE Reliability Physics Symposium, Phoenix, Arizona, April 5-7, 1983, pp. 142-147.

I-33. Ahmad, S., Blish II, R., Corbett, T., King, J., and Shirley, G., Effect of Bromine Concentration in Molding Compounds on Gold Ball Bonds to Aluminum Bonding Pads, IEEE Trans. Components, Hybrids, and Manufacturing Technology CHMT-9, pp. 379-385 (1986).
I-34. Charles, Jr., H. K., Romenesko, B. M., Wagner, G. D., Benson, R. C., and Uy, O. M., The Influence of Contamination on Aluminum-Gold Intermetallics, 20th Annual Proc. Reliability Physics Symposium, San Diego, California, March 30-31, 1982, pp. 128-139.
I-35. Khan, M. M., Tarter, T. S., and Fatemi, H., Aluminum Bond Pad Contamination by Thermal Outgassing of Organic Material From Silver-Filled Epoxy Adhesives, IEEE Trans. Components, Hybrids, and Manufacturing Technology CHMT-10, pp. 586-592 (1987).
I-36. Plunkett, P. V., and Dalporto, J. F., Low Temperature Void Formation in Gold-Aluminum Contacts, 32nd Proc. IEEE Electronic Components Conference, San Diego, California, May 10-12, 1982, pp. 421-427.
I-37. Kurtz, J., Cousens, D., and Dufour, M., Copper Wire Ball Bonding, 34th Proc. IEEE Electronic Components Conference, New Orleans, Louisiana, May 14-16, 1984, pp. 1-5.
I-38. Hirota, J., Machida, K., Okuda, T., Shimotomai, M., and Kawanaka, R., The Development of Copper Wire Bonding for Plastic Molded Semiconductor Packages, 35th Proc. IEEE Electronic Components Conference, Washington, D.C., May 20-22, 1985, pp. 116-121.
I-39. Atsumi, K., Ando, T., Kobayashi, M., and Usuda, O., Ball Bonding Technique for Copper Wire, 36th Proc. IEEE Electronic Components Conference, Seattle, Washington, May 5-7, 1986, pp. 312-317.
I-40. Levine, L., and Shaeffer, M., Copper Ball Bonding, Semiconductor International, August 1986, pp. 126-129.
I-41. Onuki, J., Koizumi, M., and Araki, I., Investigation on the Reliability of Copper Ball Bonds to Aluminum Electrodes, IEEE Trans. on Components, Hybrids, and Manufacturing Technology CHMT-10, pp. 550-555, (1987).
I-42. Riches, S. T., and Stockham, N. R., Ultrasonic Ball/Wedge Bonding of Fine Cu Wire, Proc. 6th European Microelect. Conference (ISHM), Bournemouth, England, June 3-5, 1987, pp. 27-33.
I-43. Olsen, D. R., and James, K. L., Evaluation of the Potential Reliability Effects of Ambient Atmosphere on Aluminum-Copper Bonding in Semiconductor Products, IEEE Trans. on Components, Hybrids, and Manufacturing Technology CHMT-7, pp. 357-362 (1984).

I-44. Pinnel, M. R., and Bennett, J. E., Mass Diffusion in Polycrystalline Copper/Electroplated Gold Planar Couples, Metallurgical Transactions 3, July 1972, pp. 1989-1997.
I-45. Feinstein, L. G., and Bindell, J. B., The Failure of Aged Cu-Au Thin Films by Kirkendall Porosity, Thin Solid Films 62, pp. 37-47 (1979).
I-46. Feinstein, L. G., and Pagano, R. J., Degradation of Thermocompression Bonds to Ti-Cu-Au and Ti-Cu by Thermal Aging, 29th Proc. Electronic Components Conference, Cherry Hill, New Jersey, May 14-16, 1979, pp. 346-354.
I-47. Hall, P. M., Panousis, N. T., and Menzel, P. R., Strength of Gold-Plated Copper Leads on Thin Film Circuits Under Accelerated Aging, IEEE Trans. on Parts, Hybrids, and Packaging PHP-11, No. 3, September 1975, pp. 202-205.
I-48. Pitt, V. A., and Needes, C. R. S., Thermosonic Gold Wire Bonding to Copper Conductors, IEEE Trans. on Components, Hybrids, and Manufacturing Technology CHMT-5, No. 4, December 1982, pp. 435-440.
I-49. Lang, B., and Pinamaneni, S., Thermosonic Gold-Wire Bonding to Precious-Metal-Free Copper Leadframes, 38th Proc. IEEE Electronic Components Conference, Los Angeles, California, May 9-11, 1988, pp. 546-551.
I-50. Fister, J., Breedis, J., and Winter, J., Gold Leadwire Bonding of Unplated C194, 20th Proc. IEEE Electronic Components Conference, San Diego, California, March 30-31, 1982, pp. 249-253.
I-51. James, K., Reliability Study of Wire Bonds to Silver Plated Surfaces, IEEE Trans. Parts, Hybrids and Packaging PHP-13, pp. 419-425 (1977).
I-52. Kawanobe, T., Miyamoto, K., Seino, M., and Shoji, S., Bondability of Silver Plating on IC Leadframe, 35th Proc. IEEE Electronic Components Conference, Washington, D.C., May 20-22, 1985, pp. 314-318.
I-53. Baker, J. D., Nation, B. J., Achari, A., and Waite, G. C., On the Adhesion of Palladium Silver Conductors Under Heavy Aluminum Wire Bonds, The Intl. J. for Hybrid Microelectronics 4, pp. 155-160 (1981).
I-54. Kamijo, A., and Igarashi, H., Silver Wire Ball Bonding and Its Ball/Pad Interface Characteristics, 35th Proc. IEEE Electronic Components Conference, Washington, D.C., May 20-22, 1985, pp. 91-97.
I-55. Jellison, J. L., Susceptibility of Microwelds in Hybrid Microcircuits to Corrosion Degradation, 13th Annual Proc. IEEE Reliability Physics Symposium, Las Vegas, Nevada, April 1975, pp. 70-79.

I-56. Kahkonen, H., and Syrjanen, E., Kirkendall Effect and Diffusion in the Aluminum Silver System, J. Matls. Sci. Lett. 5, p. 710 (1970).
I-57. Hermansky, V., Degradation of Thin Film Silver-Aluminum Contacts, Fifth Czech. Conference on Electronics and Physics, Czechoslovakia, October 16-19, 1972, pp. II. C-11.
I-58. Shukla, R., and Singh-Deo, J., Reliability Hazards of Silver-Aluminum Substrate Bonds in MOS Devices, 20th Annual Proc. IEEE Reliability Physics Symposium, San Diego, California, March 30-April 1, 1982, pp. 122-127.
I-59. Totta, P., Thin Films: Interdiffusion and Reactions, J. Vac. Sci. Technol. 14, 26 (1977). Also, Zahavi, J., Rotel, M., Huang, H. C. W., and Totta, P. A., Corrosion Behavior of AL-CU Alloy Thin Films in Microelectronics, Proc. of the International Congress on Metallic Corrosion, Toronto, Canada, June 3-7, 1984, pp. 311-316.
I-60. Thomas, S., and Berg, H. M., Micro-Corrosion of Al-Cu Bonding Pads, 23rd Annual Proc. Reliability Physics Symposium, Orlando, Florida, March 26-28, 1985, pp. 153-158.
I-61. Iannuzzi, M., Bias Humidity Performance and Failure Mechanisms of Non-Hermetic Aluminum SICs in an Environment Contaminated With Cl_2, 20th Annual Proc. Reliability Physics Symposium, San Diego, California, March 30-31, 1982, pp. 16-26.
I-62. Abbott, W. H., Effects of Industrial Air Pollutants on Electrical Contact Materials, IEEE Trans. on Parts, Hybrids, and Packaging PHP-10, pp. 24-27, March 1974.
I-63. Memis, I., Quasi-Hermetic Seal for IC Modules, 30th Proc. IEEE Electronic Components Conference, San Francisco, California, April 28-30, 1980, pp. 121-127.
I-64. Ahmad, S. S., Impact of Residue on Al/Si Pads on Gold Bonding, 38th Proc. IEEE Electronic Components Conference, Los Angeles, California, May 11-13, 1987, pp. 534-538.
I-65. Harman, G. G., and Leedy, K. O., An Experimental Model of the Microelectronic Ultrasonic Wire Bonding Mechanism, 10th Annual Proc. Reliability Physics Symposium, Las Vegas, Nevada, April 5-7, 1972, pp. 4956.
I-66. Harman, G. G., and Wilson, C. L., Materials Problems Affecting Reliability and Yield of Wire Bonding in VLSI Devices, Proc. 1989 Materials Research Society, Electronic Packaging Materials Science IV, Vol. 154, San Diego, California, April 24-29, 1989 (to be published).
I-67. Harman, G. G., Acoustic-Emission-Monitored Tests for TAB Inner Lead Bond Quality, IEEE Trans. on Components, Hybrids, and Manufacturing Technology CHMT-5, pp. 445-453 (1982).

CHAPTER 3

BOND FAILURES RESULTING FROM PLATING IMPURITIES AND CONDITIONS

3.1 INTRODUCTION

One of the earliest classes of documented bonding problems resulted from plating impurities in gold films. These impurities have resulted in reducing bondability (low yield), as well as causing premature bond failure during thermal stress tests or later during the life of the device (reliability). Considerable literature exists on such bond failures; however, much of it is published in plating or thin-film journals that are seldom read by packaging and wire-bonding professionals. Also, much relevant research was published a decade or more ago and is generally unavailable or is ignored. The number of plating variables is so large that there is little quantitative agreement in the literature as to the influence of a particular variable on wire bonds. Also, few experiments by different investigators are similar enough to verify previous results. Factorial statistical experiments should be run to determine the significance of each variable on bondability and reliability.

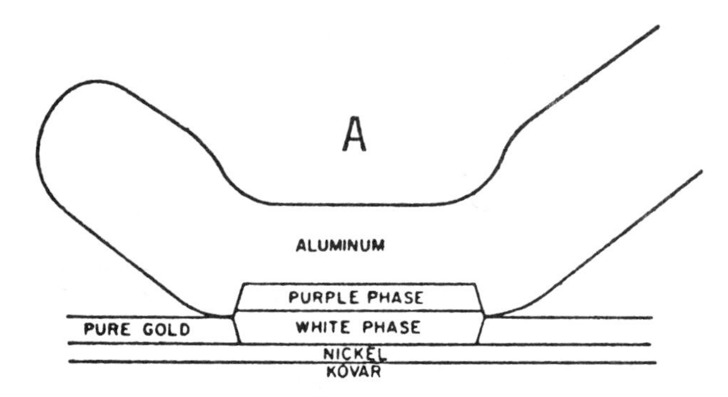

Figure P-1a. Schematic drawing of elevated temperature diffusion results for an aluminum wire bond to a pure gold plating.

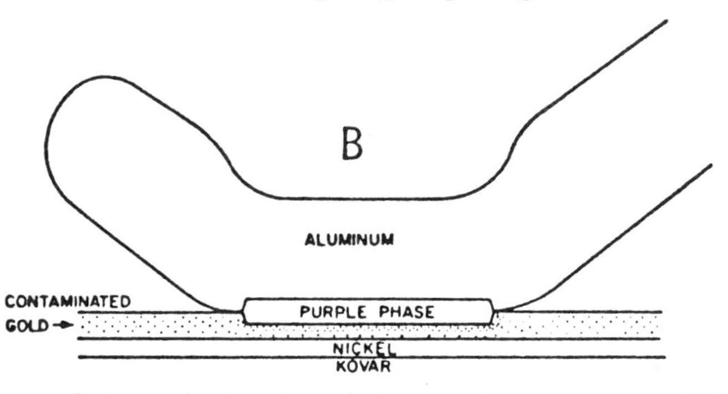

Figure P-1b. Schematic drawing of elevated temperature diffusion results for an aluminum wire bond to a contaminated gold plating (after Horsting [P-1]).

The first explanation of bond failures resulting from plating impurities was presented in a classic paper by Horsting [P-1] entitled, "Purple Plague and Gold Purity." He observed that a number of plating impurities resulted in Kirkendall-like voids and early bond failure. He hypothesized that in pure gold films the intermetallic diffusion front moved through the gold to the nickel under-plating, forming the alloy phases as shown in Figure P-1a, and the bond remains strong. However, for impure gold, the impurities are swept ahead of the intermetallic diffusion front, since most impurities in the gold have lower solubility in the intermetallic compound than in gold or aluminum, as shown in Figure P-1b. At some impurity concentration, precipitation occurs, and these particles act as sinks for vacancies produced during the diffusion reaction. Voids develop and join together, leading to weak or zero strength bonds.

The impurity analysis methods available to Horsting at that time were limited to spectrographic and chemical analysis. He was unable to identify a specific impurity that caused the problem, but found that the impure gold films contained Ni, Fe, Co, B, and other contaminants in lesser concentrations. He devised a pragmatic screening test that could detect such impure gold platings. The plated films (on Kovar headers) were multiply-bonded with aluminum wire. They were heated to 390 °C for 1 hour after which the wire bonds were pulled. If bonds lifted from the gold during the pull test, the entire header lot was rejected. He observed that well-made aluminum bonds on pure gold platings always broke in the wire, or at the bond heel after this heat treatment, the bond interface remaining strong. Horsting's theory of impurities concentrating ahead of the intermetallic diffusion front was partially verified on gold thick-films by Newsome et al [P-2] using SEM, EDAX, and Auger analysis methods.

Recently, a variation of Horsting's test (300 °C for 1 hr) has been adopted as a qualification-test method for bonds in hybrids for military usage [P-3].

3.2 SPECIFIC PLATING IMPURITIES

Gold-plating baths, intended to deposit bonding films, normally consist of potassium-gold-cyanide, plus buffers, citrates, lactates, phosphates, and carbonates in proprietary mixtures. Thallium, lead, or arsenic are added to increase plating speed and to reduce grain size. Organic "brighteners" can also be added to the bath. Thus, the problems arising in plating are very complex. They are not limited to the bath solution purity alone. It has been shown that with any given gold bath and impurity level, the deposit can vary in crystallographic structure, appearance, impurity level, hardness, hydrogen content, and density with changes in the plating waveform or current density. In addition, different plating baths, as well as bath temperatures, can produce different results at the same current density level. The film characteristics and appearance will also vary as the gold solution is depleted. Variations in these film characteristics have all been shown to influence either the bondability or the reliability of wire bonds. It is therefore not surprising that nominally the same platings obtained from different sources, or even from the same sources at different times, may cause wire-bond problems.

Thallium (Tl) was the first identified and is still the most frequently cited impurity causing bonding problems in gold platings [P-4 to P-11]. Thallium, lead [P-8 to P-10], and arsenic [P-8] are commonly added to gold-plating solutions as grain refiners to permit more rapid plating and to change the surface morphology. Thallium in gold-plated films was first identified as a source of wire-bond failures by the then new Auger electron spectrograph [P-4, 5]. It would not have been detected by wet chemical, normal spectrographic, or x-ray microprobe methods at those low concentrations. In these studies, the surface concentration of thallium was sufficient to degrade thermocompression bondability of gold wire to gold-plated lead frames.

James [P-7] found that thallium can be transferred to the gold wire from contaminated gold-plated lead frames during the crescent (second) bond break-off. He proposed that it diffused rapidly during ball formation and concentrated in the grain boundaries above the neck of the ball, where it formed a low melting eutectic. The forces and temperature applied during plastic encapsulation, or later when these devices were temperature-cycled, led to wire breaks and, thus, device failure. James showed that Tl was pervasive and could be moved around during wire bonding. However, the hypothesis that during formation of the ball, Tl diffuses through the ball during its formation into grain boundaries of the neck, and forms a very low melting-point alloy, is not proven. It is just as probable that the hydrogen torch, the bonding capillary, the lead frame bond stage clamp, or some other bonder set-up conditions were not optimum. Bonding problems often go away after almost any change in the processing schedule, and the real causes are generally not evident.

Thallium, as well as lead, in gold platings have also been shown, by Wakabayashi et al [P-10] and Evans et al [P-11], to cause premature aluminum wire-bond failures during burn-in or other heat treatment, by accelerating cracks or Kirkendall-like void formation under the bond. Evans observed such failures occurring at Tl concentrations in the plating as low as 14 ppm. Wakabayashi et al recommended that the total of all impurities in the film must be less than 50 ppm to maintain bond reliability. Figure P-2 shows that increasing the plating-current density increased the co-deposited impurity level exponentially. Thus, controlling the bath concentrations alone will not necessarily assure pure films. Endicott et al [P-8] studied the effect on bond strength of thallium, lead, and arsenic at concentrations normally used as grain refiners and compared them to platings made with pure solutions. Adding thallium and lead resulted in significant bond-strength degradation for both "as-bonded" as well as after a 150 °C thermal bake for 24 hours. However, in low-solution concentrations and low-plating-current density, adding arsenic resulted in improved bond strength under both conditions. Figure P-3 is a simplified combination of several of their figures showing these effects. This figure was chosen because it clearly compared the separate effects of all three additives. Other workers have shown more decrease in bond strength (to zero), with higher temperature or bake times [P-9,10]. It is also possible that there may be a cooperative effect between plating impurities and hydrogen in the film (see section 3.3) that results in very rapid bond-strength degradation.

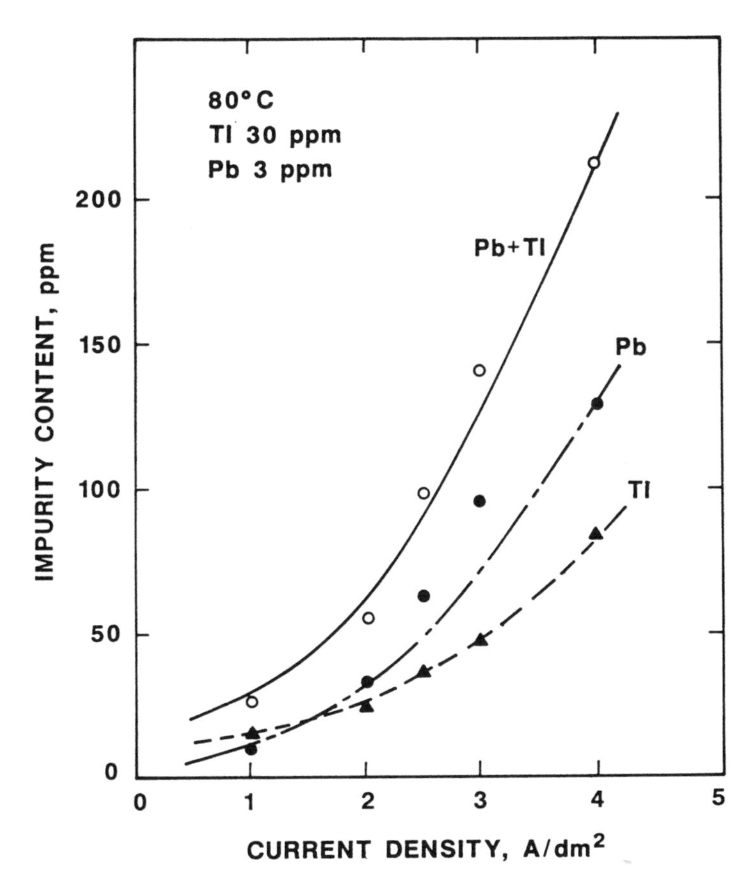

Figure P-2. Thallium and lead content in gold deposits as a function of current density. Initial bath concentration was 10 ppm for lead and 30 ppm for thallium (after Wakabayashi [P-10]).

There have also been reports that the best gold surface for bonding is the <111> crystallographic surface. For instance, Wakabayashi [P-10] states that pure gold, or gold with arsenic additives, gives primarily <111> surfaces (desirable) and that Tl and Pb additives result in an increased area of <311> surfaces. This is also complicated by the morphology effect of different plating-current densities. There are no data to definitely correlate bond quality with morphology.

3.3 Hydrogen Gas in Plated Films

Although thallium is the most frequently cited plating impurity to cause both bondability and Kirkendell-like voiding, other plating impurities or

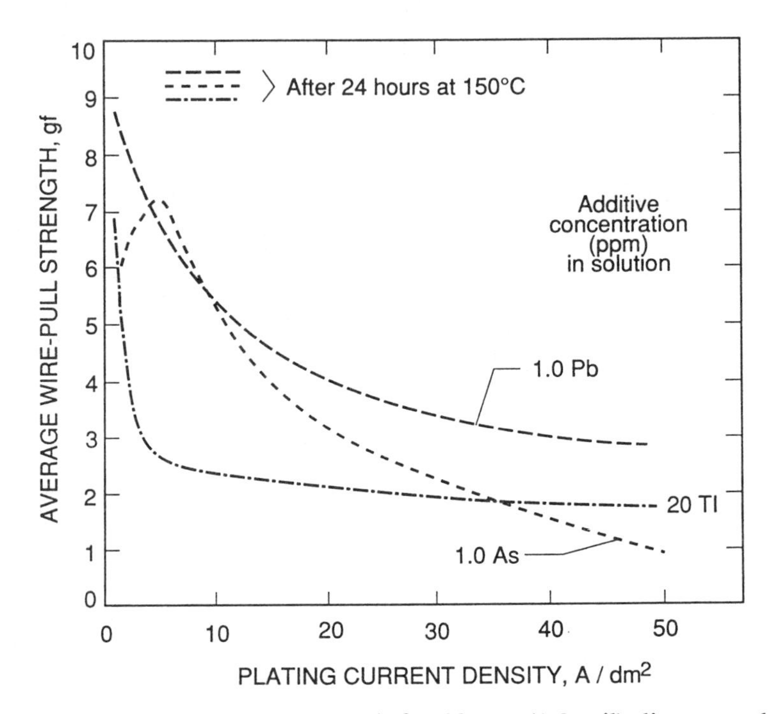

Figure P-3. Wire-bond pull strength for 32-μm (1.3-mil) diameter aluminum 1% Si wire bonds made to 1.25-μm thick plated gold films. The concentrations of grain refiners (Tl, Pb, As) are at the recommended ppm for optimum grain size (after Endicott [P-8]).

conditions have been shown to cause one or more of these same problems. Heuttner and Sanwald [P-12] were the first to investigate bondability degradation due to the presence of hydrogen bubbles in the film. These can occur as a function of plating-current density and bath-impurity level. Since they were bonding gold-coated copper wire (rather than aluminum wire) to the gold films, the bondability was addressed in their paper rather than the long-term reliability.[1] They found that the lowest bondability for plated films occurred for plating currents in the 1.6 to 2.7 A/dm^2 region, which corresponded with both the onset of rapid hydrogen evolution at the cathode and a dendritic-like (lenticular) surface morphology. Some plating conditions that can lead to increased hydrogen in the film, as well as the simplified electrochemical-plating reactions, are given in Table P-1.

[1]For strong gold-to-gold bonds, reliability is seldom a factor.

TABLE P-1
PLATING CONDITIONS THAT AFFECT HYDROGEN ENTRAPMENT

PLATING CONDITIONS THAT AFFECT HYDROGEN ENTRAPMENT

PARAMETER	CONDITION CAUSING ENTRAPMENT*
• Current density	Too high
• Agitation	Too low
• Bath gold concentration	Too low
• Bath temperature	Not critical but high is worse

• Any condition that reduces plating efficiency produces more H_2 gas at cathode and more entrapment

• Simplified anode and cathode reactions (unbalanced)

$$Au\,(CN)_2 + H_2O + e^- \rightarrow Au^0 + 1/2\ O_2\ (g\uparrow[\text{anode}]) + 1/2\ H_2\ (g\uparrow\rightarrow\text{into film}[\text{cathode}]) + 2CN^- + H^+$$

* Alec Feinberg

Gold films containing hydrogen are harder than pure films. This hardness was shown by Joshi and Sanwald [P-13] to decrease from 238 to 66 during thermal annealing at 350 °C for up to 8 hours. Much of this hardness decrease occurred in the first 10 to 15 minutes. Table P-2 gives the hardness of gold films containing hydrogen as well as their bubble-annealing characteristics. This work established the relationship between hydrogen-induced hardness and bondability on otherwise pure gold films. Joshi and Sanwald postulated that the untreated film's bondability was poor because gas entrapments caused increased ultrasonic-energy absorption, and if bubbles ruptured during bonding, the gas would prevent surfaces from coming into intimate contact. In addition, they also proposed that the gaseous entrapments might result in easy shear-force smearing of the bonding surface, and thus decreased bondability.

The effect of annealing on bondability, observed by Joshi and Sanwald, has also been observed by Kawanobe et al [P-14] for silver platings. Their platings were thinner than in reference [P-10,11] (2 to 6 μm versus 10 μm), softened more rapidly, and were fully annealed at 350 °C within 10 minutes. No gas-evolution measurements were made, and the authors were apparently unaware that this might be a factor. However, silver platings are known to incorporate hydrogen, and heat treatments are often specified. From the above, it is clear that the plating-film hardness, if not the cause of bondability problems, is at least a good indicator of them. If appropriate hardness measurement equipment is available,[2] such measurements should be made routinely on plated films. If the films are too hard, then they should be thermally annealed and hardness measurements repeated.[3]

Joshi and Sanwald concluded that during heat treatment, hydrogen-containing bubbles (which are as small as 20 Å initially) coalesce, grow, and eventually release the entrapped gas at the film's surface. They showed that the activation energy for the hydrogen-annealing process in gold is 0.35 eV for small bubbles (<500-Å diameter). This low activation energy ruled out the bubble growth by gold self-diffusion, divacancy diffusion, etc. Therefore, they proposed that the bubbles coalesced and grew by a process of structural rearrangement of the soft gold. For this mechanism, the wall between two adjacent bubbles recedes under the pressure of the heated H_2, and the two bubbles become a single, larger one.

[2]Very special hardness testers are required to make measurements on thin (typically 1.3 μm, 50 μin) gold platings used in microelectronics. These require loads approximately 1-2 grams. Sometimes platings are made much thicker so that more conventional hardness equipment can be used.

[3]Keep in mind that nickel, copper, chromium, or other non-noble undercoats may be diffused to the surface, and then oxidize, and themselves cause bondability and reliability problems.

TABLE P-2
TIME-DEPENDENT CHANGES ON ANNEALING 10-μm THICK GOLD PLATINGS CONTAINING HYDROGEN ENTRAPMENTS AT 350 °C

Anneal. Time (sec)	Bubble Density (N_v/cm^3)	Knoop Hardness 2gm load (kg/mm^2)	Resistivity p, at 78K (μΩ-cm)	H_2 Gas Released	
				(cm^3/g Au)	(cm^3/cm^3 Au)
0	8.5×10^{16}	238	1.25	0	0
60	—	203	0.93	—	—
600	6.7×10^{16}	134	0.72	—	—
6,000	8.7×10^{15}	97	0.60	—	—
30,000	6.4×10^{15}	66	0.55	0.15	28.9

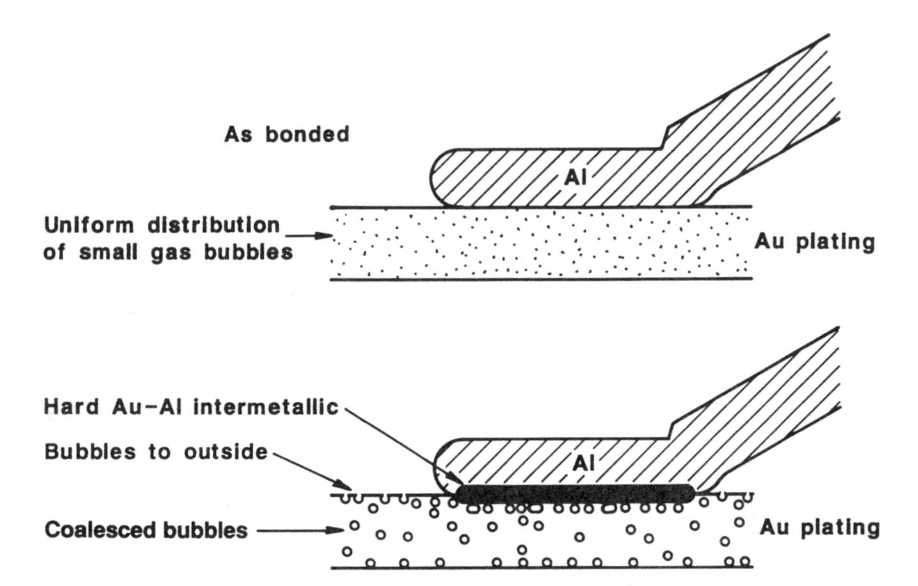

Figure P-4. Proposed failure mechanism for aluminum wire bonds on plated gold films containing hydrogen gas entrapments: (top) as bonded, (bottom) after thermal stress (after Harman 1988 ISHM short course).

Gold films containing gas have been observed to severely decrease long-term reliability when used for aluminum wire bonding,[4] and the structural rearrangement mechanism can be used to explain these failures. After a thermal stress, an aluminum bond on a gold film will develop an interfacial layer of Au-Al intermetallic compound. The activation energies of structural rearrangements in very hard, brittle, gold-aluminum, intermetallic compounds are believed to be much higher than in gold (probably higher than 1 eV, if these rearrangements exist at all), although such measurements have not been made. With these assumptions, bubbles could not move into the intermetallic region and would pile up at the interface, as shown in Figure P-4. (Note the similarity between this and Horsting's model, shown in Figure P-1.) During a subsequent failure analysis, sectioning of the bonds would reveal apparent ''Kirkendall'' voids, which would be an ***incorrect*** diagnosis. Additional analysis of the film by Auger or SIMS would not normally detect hydrogen gas. One wonders how many similar bonding problems have been misdiagnosed in the past. It is possible that some of Horsting's [P-1] gold platings failed because of H_2 gas or, at least, through a synergism of this mechanism with other impurities. His

[4]Alec Feinberg, private communication.

TABLE P-3*
PRACTICAL ANNEALING SCHEDULE FOR HYDROGEN CONTAINING GOLD FILMS

Temp.	Annealing time
150 °C	50 hr
175	29
200	18
250	8
300	4
350	2.25
390	1.5

*time = 0.0033 exp [4062/T°K]

pragmatic screening procedure (390 °C for 1 hr - very similar to the annealing time in Table P-3) would not distinguish between one type of void and another. It should be stated that a gold-wire bond on a H_2-filled plating would not fail under the above conditions since no pile-up of bubbles would occur, because they could begin to move into the gold wire or ball.

Figure P-5a,b is a photomicrograph of a gold plating that had resulted in wire-bond failures of the type shown in Figure P-4. The film shown in Figure P-5b was heated for 4 hours at 300 °C to reveal gas-entrapment craters on its surface. Control platings without gas entrapment showed little change in surface appearance. Lower baking temperatures for longer periods will produce the same effect as shown in Figure P-5b. Table P-3 gives practical hydrogen-annealing times for gold-plated films calculated from the activation energy of 0.35 eV (from Joshi and Sanwald).

Free hydrogen inside a package has been shown to inhibit the formation of Au-Al compounds by filling vacancies in the aluminum and its grain boundaries [P-15], inhibiting aluminum diffusion. The proposed hydrogen-gas bubble mechanism would be little affected by free hydrogen since all aluminum in the gold film would already be in intermetallic compound form.

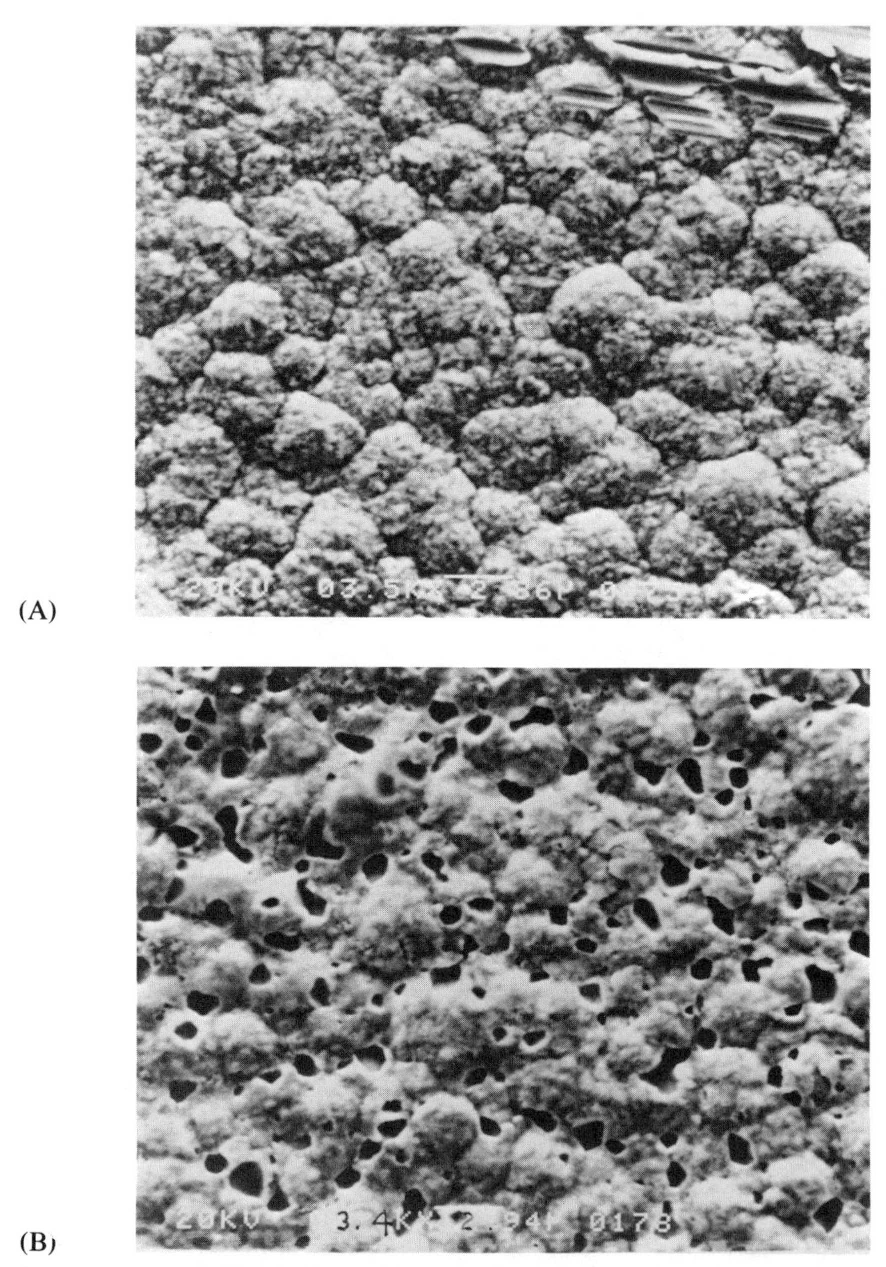

Figure P-5. Plated gold films with high levels of gaseous entrapment. (A) High entrapment—is plated, (B) same as (A) but heat treated 4 hours at 300 °C. Bars on photos approximately 2.8 μm long.

3.4 HYDROGEN-INDUCED PACKAGE PROBLEMS

There may be other types of failures that result from hydrogen entrapment in various parts of electronic packages. Many package substrate materials, such as iron-nickel alloys or a Ni strike (under plating), have a high solubility for H_2, and this may relieve the bubble pile-up effect at that lower interface. It would also embrittle these materials and lead to other types of failures.[5] Hydrogen-filled gold platings on relatively hard copper-lead frames (which have a low-hydrogen solubility) would presumably show a tendency to delaminate, in the manner illustrated in Figure P-4.

3.5 PLATING STANDARDS

It should not be surprising that gold platings can contain hydrogen or that heat treatment can remove it. For instance, high-strength plated steels are routinely heat-treated at about 200 °C for up to 24 hours to purge hydrogen and thus prevent embrittlement, and some platings (e.g., cadmium) are worse than others in regard to this hydrogen problem. Many plating standards [e.g., Fed. Spec. QQ-S-365-C, AMS(SAE)2422B, etc.] contain requirements for heat-treating plated films to eliminate hydrogen, and such treatment should be included for gold films, at least if they are to be used for bonding. There are no generally accepted standards for platings used for wire bonding, and Horsting's [P-1] pragmatic screen, or equivalent, is still necessary. The military plating specifications for electronics (MIL-G-45204) does not include adequate film purity, hardness, plating-current density criteria, or hydrogen content. These specifications are worthless for assuring high-quality wire bonds and are more appropriate for connector contacts.

3.6 RECOMMENDATIONS FOR RELIABLE GOLD-PLATED FILMS

Gold films with desirable characteristics (bondability, reliability) should, based upon the foregoing, contain no measurable thallium and less than 50 ppm total of Ni, Cu, and Pb impurities. Analysis should be made on a sample basis on the film itself if possible,[6] since many variables can change

[5]Gold platings over nickel strikes on lead frames at times show a tendency to crack and delaminate at bends. This is usually caused by improper nickel plating conditions or thickness. However, hydrogen embrittlement from any plating process as described above could cause or enhance this problem. As in the gold film, normal Auger or other analyses would not detect the presence of hydrogen in the film and the cause of these failures would remain a mystery.

[6]Continued analysis of plating lots to 50 ppm would be prohibitively expensive. Perhaps a pragmatic test, such as Horstings [P-1] bake test, should be applied to lots and only failed lots analyzed.

the correlation between plating solution and film impurity level. Hydrogen and other gas occlusions should be to a minimum. The film should be soft with a hardness of 40 to 70 HKN[2] and nodular in appearance. It should not be shiny nor have a lenticular surface structure. It is possible that a <111> dominant-surface structure is desirable, but this has not been proven. Unbondable or unreliable gold films that contain hydrogen can be hardness tested, annealed (see Table P-3), hardness tested again (if equipment is available[2]), and if softer, should be usable, unless the gold is plated over nickel, chrome, titanium, or copper. During annealing, these metals may diffuse to the surface, oxidize, and render the gold unbondable. This oxide must then be chemically removed if the plating is to be bonded. Gold platings over combination adhesion-barrier coatings, such as titanium-palladium, limit diffusion and can generally be heat treated and used without further cleaning. Since plating conditions such as waveform, bath temperature, and composition can affect the film characteristics, there is no specific recommendation for a plating current density to achieve bondable films. The literature is contradictory in this regard and more work is needed. If grain refiners or Ni, Cu, etc. are in the bath, then high plating rates will generally increase their incorporation in the films and produce bonding problems.

3.7 METALLIC IMPURITIES IN OR ON GOLD FILMS THAT ARE NOT AN INTENTIONAL PART OF PLATING BATHS

There are numerous metallic contaminants that can be incorporated into the bond interface and degrade either bondability or reliability. Some may be accidentally introduced into a plating bath, others may be diffused up from the substrate, and still others may occur as a result of some later chemical or "cleaning" step and remain on the surface during bonding. These contaminants will be discussed along with any known cleaning techniques. One must be aware, however, that in a dynamic technology, contaminants can be introduced from a source that is unanticipated and unknown at the time of this writing. Most metallic contaminants that affect bonds appear as surface films of from 20 to 200 Å thick. If they consist of non-noble metals, they will generally be oxidized by various heat or chemical treatments before bonding, and the oxide may lower bondability (especially for TC bonding) and sometimes lower reliability. Hard-brittle oxides that occur on soft metals (e.g., Al_20_3 on Al) break up and are pushed into "debris zones" during bonding as the aluminum deforms. These generally have little effect on the bonding process, but hard oxides on hard metals (e.g. TiO_2 on Ti) will decrease bondability. Softer oxides (e.g., copper and nickel oxides) are observed to decrease bondability (increase the activation energy required to form a bond), possibly by serving as a lubricant in the bond interface.

Often, failure analyses reported in the literature may reveal contaminants in the interface of a failed bond, without being able to determine the source or even which steps in the packaging process were responsible for their introduction. In other reported cases, it was not clear that any one indicated impurity actually caused the failure. A complete study may have shown that a poor bonding machine setup was as much a cause of the failure as the various contaminants revealed in the failure analysis. However, the metal contaminants discussed below (Ni, Cu, Cr, Ti, Sn) are believed to be adequately documented as to their effect on bonding.

After thallium, ***nickel*** is the most frequently cited material to degrade bonds on gold-plated films [P-1,16-21]. It is generally considered to affect bondability but is also cited in [P-1] as affecting reliability. Nickel may be introduced into a gold-plating bath by some accident, such as a Kovar (iron, nickel, cobalt) lead frame falling into the bath and slowly dissolving. Another mechanism for nickel to enter the gold film is by thermally diffusing upward from a thin strike plating. It may move rapidly via grain boundary diffusion [P-18] to the surface of the gold during high-temperature die attachment or other heat treatment. There, it spreads over the gold surface by surface diffusion [P-21], oxidizes, and then renders the surface unbondable. The effect of nickel on TC bondability is shown in Figure P-6. As with thallium, nickel and copper concentrations in the gold film are strongly dependent on the plating-current density, as shown in Figure P-7.

Copper [P-17,18,20,22-25] from plating-bath contamination, lead frames, etc. can follow the same diffusion route as nickel and will also decrease the bondability. Various authors disagree which impurity, Cu [P-17] or Ni [P-20], has the most effect on bond-strength degradation, and this controversy may be related to the analysis method, the surface concentration, the impurity-plating rate (Figure P-7), the solution concentration, etc. If measured as atomic percent on the gold surface, then nickel is worse [P-20]. Both Cu and Ni impurities should be avoided since they readily oxidize, and the oxides degrade bondability. Both copper and nickel are still used as lead frames or as package platings and are often bonded directly. Ultrasonic, TC, or TS bonding can be done with a high yield and maintain reliability only if the surfaces are oxide free. Thus, for such bonding, an appropriate method of preventing oxide growth during storage or a chemical removal process is essential. It may be helpful in bonding directly to pure copper or nickel to apply ultrasonic energy as the wire initially contacts the metal. This helps clear the oxide away before microwelds form and wire-to-substrate motion ceases.

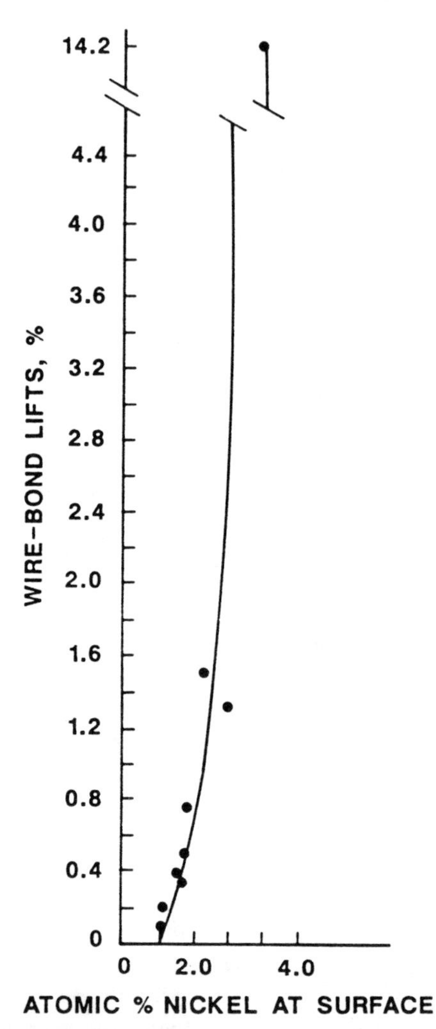

Figure P-6. Wire-bond lifts versus Auger atomic percent nickel on surface of gold plating. Thermocompression wedge bonds of 38-μm (1.5-mil) diameter gold wire (after Casey [P-16]).

Chromium, which is often used to promote adhesion between substrates and vacuum-deposited gold films, can rapidly diffuse through grain boundaries to the surface and oxidize [P-23,26-29]. Figure P-8 illustrates grain boundary diffusion for chromium through gold (as well as copper, nickel, and other metals). Heating chromium-gold films for two hours at 250 °C was observed to impair the thermocompression bondability of 3-μm thick

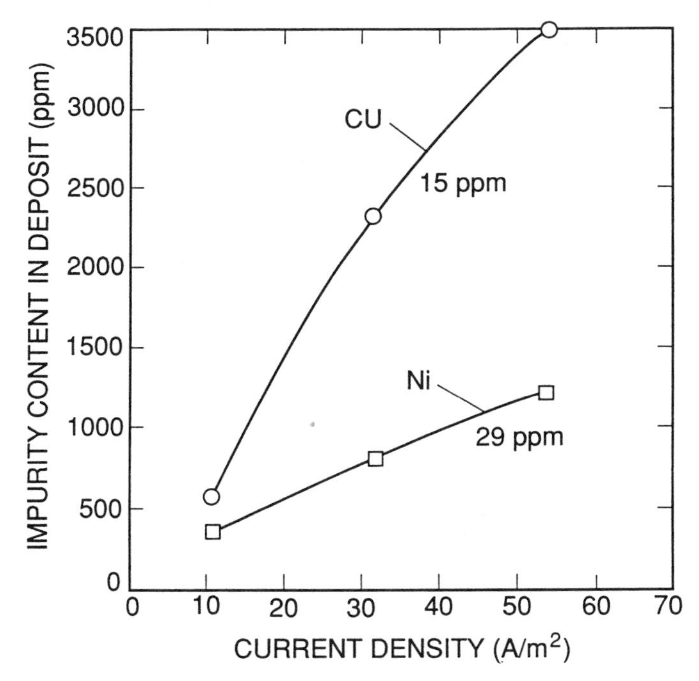

Figure P-7. Influence of plating current density on the impurity content of gold deposits (after Dini and Johnson [P-20]).

gold films [P-26]. Higher temperatures or thinner films can decrease the diffusion time significantly. A special cleaning etch, cerric ammonium nitrate, was developed by Holloway and Long [P-28] to remove the chromium oxide from gold surfaces and completely restore thermocompression bondability.

Titanium, as with chromium, is used in various substrate-metallization systems and can diffuse up [P-30] to the gold surface or deposit on it during some part of the processing procedure [P-31]. It will oxidize and decrease bondability. Thompson et al [P-31] used a dilute 10:1 HF, $HN0_3$ etch on the gold surface to restore bondability. Note, however, that any remaining traces of this etch could degrade the reliability of aluminum wire bonds on the surface, although gold bonds would remain unaffected.

Tin has been identified in two studies as leading to bond failures. Vaughan and Raut [P-32] found that tin was deposited (possibly as the oxide - from Auger analysis) on the gold substrate while using a contaminated cleaning solution. Tin thickness in the range of 20 to 30 Å reduced bondability significantly. A potassium carbonate rinse was used to remove the tin from the substrates. This rinse left traces of potassium on the surface which

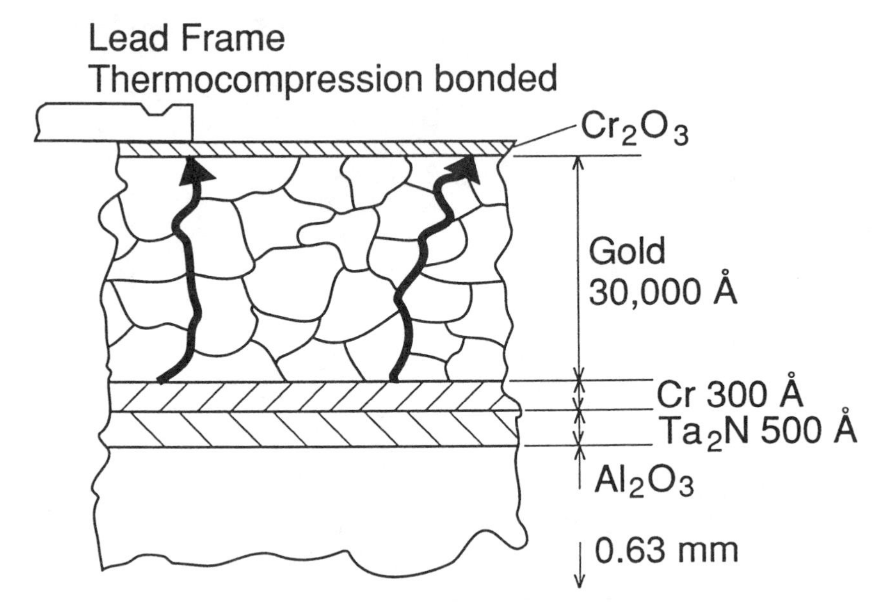

Figure P-8. Hybrid microcircuit geometry with Cr_2O_3 layer on the surface. The arrows indicate possible grain boundary diffusion paths (after Nelson and Holloway [P-29]).

could affect some active devices. Tin was also cited by Davis and Joshi [P-33] as causing gold thermocompression-bond failures to aluminum pads. Apparently, it prevented a strong bond from forming. This work was not clear as to the sources or the extent of the tin problem, since carbon that could have been a contributing factor was also found in the interface. Nevertheless, upon aging, tin does form an oxide which is tenacious and impervious enough to reduce solderability in tin-plated lead frames, so it should also be considered a potential hazard for wire bonding.

3.8 REFERENCES

P-1. Horsting, C., Purple Plague and Gold Purity, 10th Annual Proc. IRPS, Las Vegas, Nevada, April 5-7, 1972, pp. 155-158.

P-2. Newsome, J. L., Oswald, R. G., and Rodrigues de Miranda, W. R., Metallurgical Aspects of Aluminum Wire Bonds to Gold Metallization, 14th Annual Proc. Rel. Phys., Las Vegas, Nevada, April 20-22, 1976, pp. 63-74.

P-3. Military Standard 883, Test Methods and Procedures for Microelectronics, Method 5008-Test Procedures for Hybrid and Multichip Microcircuits.

P-4. McDonald, N. C., and Palmberg, P. W., Application of Auger Electron Spectroscopy for Semiconductor Technology, Intl. Electron Devices Meeting, Washington, D.C., October 11-13, 1971, p. 42.

P-5. McDonald, N. C., and Riach, G. E., Thin Film Analysis for Process Evaluation, Electronic Packaging and Production, April 1973, pp. 50-56.

P-6. Czanderna, A. W., (Ed.), Methods of Surface Analysis VI, Chapter 5, pp. 212-222., Elsevier Scientific Publishing Co., New York, New York (1975).

P-7. James, H. K., Resolution of the Gold Wire Grain Growth Failure Mechanism in Plastic Encapsulated Microelectronic Devices, IEEE Trans. on Components, Hybrids, and Manufacturing Technology CHMT-3, September 1980, pp. 370-374.

P-8. Endicott, D. W., James, H. K., and Nobel, F., Effects of Gold-Plating Additives on Semiconductor Wire Bonding, Plating and Surface Finishing V, November 1981, pp. 58-61.
P-9. Okumara, K., Degradation of Bonding Strength (Al Wire - Au Film), by Kirkendall Voids, J. Electrochem. Soc. 128, pp. 571-575, 1981.
P-10. Wakabayashi, S., Murata, A., and Wakobauashi, N., Effects of Grain Refiners in Gold Deposits on Aluminum Wire-Bond Reliability, Plating and Surface Finishing V, August 1982, pp. 63-68.
P-11. Evans, K. L., Guthrie, T. T., and Hays, R. G., Investigation of the Effect of Thallium on Gold/Aluminum Wire Bond Reliability, Proc. ISTFA, Los Angeles, California, 1984, pp. 1-10.
P-12. Huettner, D. J., and Sanwald, R. C., The Effect of Cyanide Electrolysis Products on the Morphology and Ultrasonic Bondability of Gold, Plating and Surface Finishing, August 1972, pp. 750-755.
P-13. Joshi, K. C., Sanwald, R. C., and Annealing, H., Behavior of Electro-deposited Gold Containing Entrapments, J. Electronic Materials 2, pp. 533-551, 1973.
P-14. Kawanobe, T., Miyamoto, K., Seino, M., and Shoji, S., Bondability of Silver Plating on IC Leadframe, Proc. 20th IEEE Electronic Components Conference, Washington, D.C., pp. 314-318, May 20-22, 1985.
P-15. Shih, D-Y., and Ficalora, P. J., The Reduction of Au-Al Intermetallic Formation and Electromigration in Hydrogen Environments, 16th Annual Proc. IRPS 1978, San Diego, California, April 18-20, 1978.
P-16. Casey, G. J., and Endicott, D. W., Control of Surface Quality of Gold Electrodeposits Utilizing Auger Elec. Spec., Plating and Surface Finishing, July 1980, pp. 39-42.
P-17. McGuire, G. E., Jones, J. V., and Dowell, H. J., Auger Analysis of Contaminants that Influence the Thermocompression Bonding of Gold, Thin Solid Films 45, pp. 59-68, (1977).
P-18. Hall, P. M., and Morabito, J. M., Diffusion Problems in Microelectronics Packaging, Thin Solid Films 53, pp. 175-182 (1978).
P-19. Endicott, D. W., and Casey, G. J., High Speed Gold Plating from Dilute Electrolytes, Proceedings American Electroplaters Society, paper 1-d3, 1979.
P-20. Dini, J. W., and Johnson, H. R., Influence of Codeposited Impurities on Thermocompression Bonding of Electroplated Gold, Proc. ISHM Symposium 1979, pp. 89-95, Los Angeles, California, October.
P-21. Loo, M. C., and Su, K., Attach of Large Dice with Big Glass in Multilayer Packages, Hybrid Circuits (UK) Number 11, pp. 8-11, September 1986.

P-22. Panousis, N. T., Thermocompression Bondability of Bare Copper Leads, IEEE Trans. on Components, Hybrids, and Manufacturing Technology CHMT-1, pp. 372-376, 1978.
P-23. Panousis, N. T., and Hall, P. M., Application of Grain Bonding Diffusion Studies to Soldering and Thermocompression Bonding, Thin Solid Films 53, pp. 183-191 (1978).
P-24. Dini, J. W., and Johnson, H. R., Optimization of Gold Plating for Hybrid Microcircuits, Plating and Surface Finishing, January 1980, pp. 53-57.
P-25. Spencer, T. H., Thermocompression Bond Kinetics—-The Four Variables, Intl. J. Hybrid Microelectronics 5, pp. 404-408 (1982).
P-26. Panousis, N. T., and Bonham, H. B., Bonding Degradation in Tantalum Nitride-Chromium Gold Metallization System, 11th Annual Proc. Reliability Physics, pp. 21-25, Las Vegas, Nevada, April 3-5, 1973.
P-27. Harman, G. G., The Use of Acoustic Emission in a Test for Beam-Lead, TAB, and Hybrid Chip Capacitor Bond Integrity, IEEE Trans. on Parts, Hybrids, and Packaging PHP-13, pp. 116-127 (1977).
P-28. Holloway, P. H., and Long, R. L., On Chemical Cleaning for Thermocompression Bonding, IEEE Trans. on Parts, Hybrids and Packaging, PHP-11, pp. 83-88, (1975).
P-29. Nelson, G. C., and Holloway, P. H., Determination of the Low Temperature Diffusion of Chromium Through Gold Films by Ion Scattering Spectroscopy and Auger Electron Spectroscopy, ASTM Special Technical Publication 596, Surface Analysis Techniques, pp. 68-77 (1976).
P-30. Donya, A., Watari, T., Tamura, T., and Murano, H., GLO: A New Technology for Fabrication of Fine Lines on Multilayer Substrate, Proc. IEEE Electronics Components Conference, Orlando, Florida, pp. 304-313 (1983).
P-31. Thompson, R. J., Cropper, D. R., and Whitaker, B. W., Bondability Problems Associated with the Ti-Pt-Au Metallization of Hybrid Microwave Thin Film Circuits, IEEE Trans. on Components, Hybrids, and Manufacturing Technology CHMT-4, pp. 439-445, (1981).
P-32. Vaughan, J. G., and Raut, M. K., Tin Contamination During Surface Cleaning for Thermocompression Bonding, Proc. ISHM, 1984, pp. 424-247.
P-33. Davis, L. E., and Joshi, A., Analysis of Bond and Interfaces with Auger Electron Spectroscopy, Proc. Advance Techniques in Failure Analysis, Los Angeles, California, October 1977, pp. 246-250.

CHAPTER 4

CLEANING TO IMPROVE BONDABILITY AND RELIABILITY

4.1 CLEANING METHODS

4.1.1 Introduction

Various cleaning methods have long been used to remove contaminants at different stages of wafer processing. However, until recently there was little consideration given to a cleaning step specifically designed to improve the yield and reliability of wire bonds. Modern VLSI devices are complex, can have hundreds of I/Os, and may have reliability requirements unheard of ten or fifteen years ago. Modern bond pad metallizations are often harder than in the past, containing various additives, and in addition, reactive ion processing of the wafer can leave fluorocarbon films on the surface. All of these can inhibit bondability and affect reliability. Because of the extensive handling, as well as the use of polymer die attach, the hybrid industry was the first to adapt molecular cleaning methods before bonding. Their use has recently expanded into some other packaging areas.

Contaminants on bond pads have long been known to degrade both the bondability and the reliability of wire bonds. Table CL-1 is a reasonably complete list of the contaminants that have been found to degrade bonds. The table is only indicative since the effect on bonds may be concentration

TABLE CL-1
IMPURITIES THAT LEAD TO WEAK BONDS

- **Halogens from**
  - Plasma etching (dry processing)
  - Epoxy outgassing
  - Silox etch
  - Photoresist stripper
  - Solvents (TCA, TCE, chloro-fluro's)

- **Contaminants from plating**
  - Thallium
  - Brighteners
  - Lead
  - Iron
  - Chromium
  - Copper
  - Nickel
  - Hydrogen

- **Sulfur from**
  - Packing containers
  - Ambient air
  - Cardboard & paper
  - Rubber bands

- **Miscellaneous organic contaminants from**
  - Epoxy outgassing
  - Photoresist
  - General ambient air (poor storage)
  - Spittle

- **Others that cause corrosion or inhibit bonding**
  - Sodium
  - Chromium
  - Phosphorous
  - Bismuth, cadmium
  - Moisture
  - Glass, vapox, nitride
  - Carbon
  - Silver
  - Copper
  - Tin

dependent or may only act synergistically, for example, with water vapor or heat, or in Au-Al interfaces. Some contaminants primarily affect bondability while others reduce reliability. The most important contaminants that affect bond quality are discussed elsewhere in this monograph.

There are many human sources of contamination not listed in Table CL-1 that may contain bond-degrading contaminants. Some of these are small particles of skin, hair, sweat, spittle, and mucus. These may arrive at the device surface by the driving force of talking, coughing, sneezing, yawning, head shaking, scratching, etc. A compilation of human sources of contamination was given by Thomas and Calabrese [CL-1]. A person sitting motionless generates about 10^5 particles per minute of greater than 0.3-μm diameter and up to 50 times more particles while moving. A fully suited person, walking in a class 100 clean room, will distribute 50,000 particles in that same period of time [CL-2]. Other sources of contamination may enter the air from drinking water (Cl and Br) or from dry cleaned clothes (tetrachloroethylene) [CL-3].

Considering the large number of possible bond-degrading contaminants, a variety of methods could be required to clean surfaces containing several contaminants. Some of these contaminants (e.g., halogens) can become chemically bound to bonding pads and require treatments that can only be performed at the wafer level, such as heating in oxygen for 30 minutes at 350 °C [CL-4]. Others, such as glass, nitride, and some metal oxides on pads, also cannot be removed at the packaging level. Organics, however, may be easily removed with UV-ozone or O_2 plasma after die attach, immediately before bonding.

This section will present evidence that both bondability and reliability can be affected by contamination, and that some assembly line cleaning methods can remove that contamination. Two molecular cleaning methods that are the most successful for cleaning die-attached chips before bonding will be primarily discussed. Various solvent techniques [CL-5 to 7] (solution, vapor-phase fluorocarbons, ionographic, and DI water) will be discussed only as they compare with the gaseous cleaning methods.

4.1.2 Molecular Cleaning Methods

Both plasma and UV-ozone cleaning methods have been known for many years [CL-7 to 15]. Sowell et al [CL-8] gave the clearest comparison of UV-ozone,[1] argon plasma, and ultra-high-vacuum bake-out methods of

[1]The combination of radiation from their high-pressure mercury, quartz tube lamp, and the low-pressure oxygen environment would only produce a small amount of ozone. Therefore, the long cleaning times (2 hours) appears to be nearer the times found by Vig [CL-12] for cleaning by UV alone.

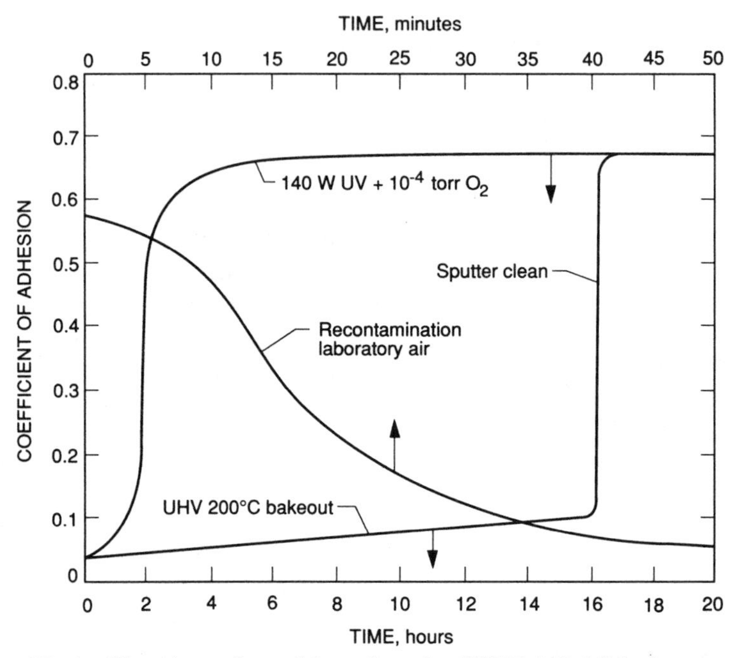

Figure CL-1. Cleaning of a gold surface by UHV 200 °C bakeout, argon sputtering, and UV irradiation at 10^{-4} Torr O_2 (lower scale). Recontamination rate in normal laboratory air is also shown (upper scale) (after Sowell et al [CL-8]).

cleaning gold surfaces. Their data are reproduced in Figure CL-1. The coefficient of adhesion (related to the coefficient of friction) for gold in vacuum is used as the measure of a clean surface. Recontamination by hydrocarbons in laboratory air is indicated by the center curve. These data correlated to data measured for glass surfaces, where the water-drop contact-angle method of evaluation was used. Much of the classical work in cleaning and contamination control is presented in a book edited by Mittal [CL-9] and should be referred to for more detailed fundamental information. The remainder of this section will describe work directly applied to the bondability and reliability of wire bonds.

Ultraviolet-Ozone Cleaning

Ultraviolet-ozone cleaners generally consist of a chamber containing banks of quartz-envelope, low-pressure, mercury vapor lamps. These are designed to emit significant amounts of radiation of 1849-Å and 2537-Å wavelengths. Devices to be cleaned are placed in the chamber as close as practical to the lamps. Since ozone gas is considered dangerous, the units are usually operated in a fume hood or at least where there is some means of removing the gas from the area.

The removal of organic contaminants with ultraviolet-ozone takes place as follows. The 1849-Å UV energy breaks up the O_2 molecule into atomic oxygen (O + O) which combines with other O_2 molecules to form ozone, O_3. Ozone has a strong absorption for 2537-Å UV and may break up again into atomic oxygen and O_2. Any water present may also be broken into the OH^- free radical. All of these (OH, O_3, and O) can react with hydrocarbons to form CO_2 + H_2O which leave the device surface as a gas. The strong 2537-Å UV may additionally break the chemical bonds of the hydrocarbon, accelerating the oxidation process. Early work by Jellison [CL-10] showed that UV (2537 Å) alone could clean gold metallization of carbonaceous films and increase thermocompression ball-bond shear-strength.

Figure CL-2 shows an example of such cleaning to increase bondability. Even a few-angstroms-thick film of carbon[2] was found to impair bondability, whereas a cleaned gold film (<1-Å carbon) can be strongly TC bonded at 150 °C, which is the minimum temperature generally used in thermosonic bonding! Holloway and Bushmire [CL-11] found similar cleaning results with ozone alone (the UV creating it was shielded from the samples). However, Vig and Le Bus [CL-12] found that UV and ozone (2537 Å + 1849 Å + ozone) cleaned much faster than UV or ozone alone, up to 100 times faster depending on the specific impurities. Therefore, present cleaners employ the combination.

Jellison compared the bondability of gold thick-films that were cleaned by vapor degreasing and boiling trichlorethylene to those cleaned with UV-ozone. His films were contaminated with beeswax, petrolatum, and halocarbon wax. The results indicated that vapor degreasing was a poor cleaning procedure for removing beeswax (see Figure CL-3), but quite effective for petrolatum and halocarbon wax. Ultraviolet-ozone, however, effectively removed all of the contaminants. This points out the major problem of solvent cleaning. No one solvent is apt to remove all organic contaminants that may be on a bond pad, emphasizing the importance of a molecular cleaning method.

Some general aspects of UV-ozone cleaning have been reviewed by the equipment manufacturers. Clarke [CL-14] described applications to hybrid circuit cleaning including such components as surface-acoustic-wave (SAW)

[2]The actual thickness of a carbon containing film may be 3 to 4 times the measured carbon thickness.

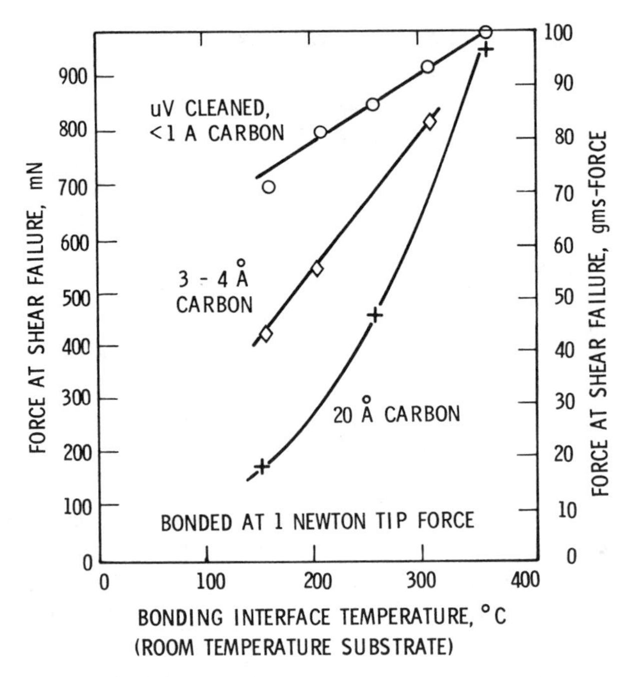

Figure CL-2. Effect of surface contamination on the thermocompression bonding of gold, pulsed bonding (after Jellison [CL-10]).

devices. Zafonte and Chiu [CL-15] described its application to the cleaning of silicon wafers.

4.1.3 Plasma Cleaning

Plasma-cleaning equipment is generally larger, more costly, and more complicated than UV-ozone equipment. It requires a vacuum pump, a several-hundred-watt RF power generator, and pure gasses (usually oxygen and argon). By its nature it is a batch cleaning method, whereas UV-ozone could, in principle, be a belt-driven in-line system. In use, devices are placed in a chamber, which is evacuated, the appropriate gas is introduced (approximately 0.1 Torr), and RF power is switched on for from about 10 to 30 minutes to effect the cleaning process. The earliest use of plasma cleaning in microelectronics was to remove photoresist contamination from wafers [CL-16]. However, more recently there have been numerous studies applying plasma (O_2 and/or Ar) to the removal of contaminants from bond pads or from hybrid substrates [CL-16 to 28]. Bonham [CL-18] was the first to use oxygen plasma cleaning on hybrid microcir-

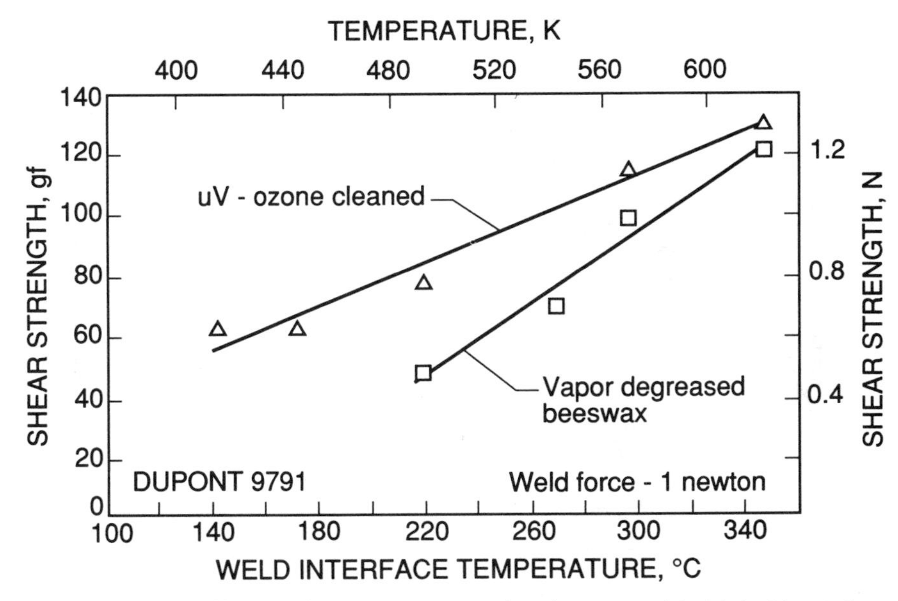

Figure CL-3. Effect of beeswax contamination on gold thick film (after Jellison [CL-13]).

cuits. He found that such cleaning improved the bondability and reliability of gold-wire bonds to aluminum pads on epoxy die-attached devices. An example of the increased reliability after such cleaning is shown in Figure CL-4. Similar improvements in bondability were obtained by White [CL-19] for gold-ball bonding to gold-plated surfaces. He used oxygen plasma to remove die attach epoxy "bleed" from bond pads on substrate metallization near the chip. White also showed that such plasma cleaning had no negative effect on die-shear strength. These results were verified by Kenison et al [CL-22] and Buckles [CL-23]. Khan et al [CL-28] used O_2/N_2 plasma to effectively remove epoxy thermal outgas material from bond pads.

Graves [CL-20] evaluated various plasma processes for improving the bondability of hybrids. He found that the bondability of the particular thick-film gold (Dupont 4290, reaction-bonded) used in their production was not improved by oxygen plasma, possibly because of oxidation of the reaction bonding elements, such as copper. His studies showed that the best results were obtained with oxygen-free argon plasma (0.25 Torr, 300 W, 60 min[3]).This work was also interesting because Graves found that

[3]Note that this cleaning time and power are on the very high side of typical plasma-cleaning parameters.

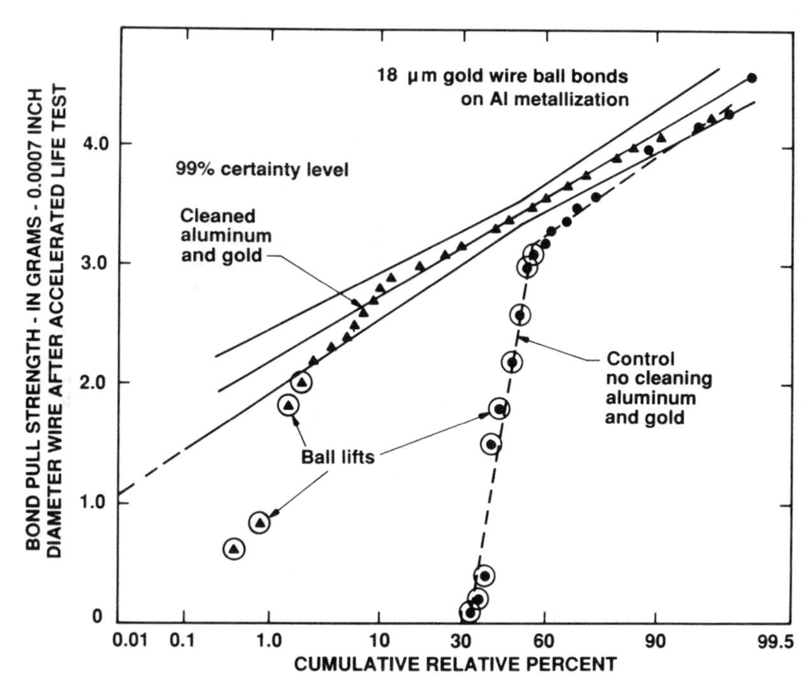

Figure CL-4. Influence of O_2 plasma cleaning on reliability of gold bonds on aluminum metallizations after a thermal stress of 300 °C for 4 hours (after Bonham and Plunkett [CL-18]).

optimum results depended not only upon the gas and RF power, but also upon the fixturing, as well as the specific material being cleaned. Presumably, fixturing can shield or otherwise change the concentration of ionized plasma in a local region. In another study, Graves [CL-21] found that an oxygen plasma did ***not*** remove fluorine contamination from semiconductor bond pads. Presumably, the fluorine had chemically reacted with the aluminum under the surface oxide.

The mechanism of oxygen plasma cleaning is similar to that of UV-ozone. Some of the O_2 can become ionized and other O_2 breaks apart into atomic oxygen, O + O. These react with the hydrocarbons to form H_2O and CO_2 [CL-27]. There is also energetic bombardment by the excited oxygen atoms, which assists in breaking up the hydrocarbon molecules, as well as in sputtering off the contaminants. Ionized argon is not known to form stable compounds, although it may form brief metastable compounds with carbon or other contaminants, removing them and then decomposing, releasing them to be pumped out of the gaseous plasma. Argon has more than twice the atomic weight of oxygen, and it can knock off various forms of contamination by an impact (sputtering). In general, it takes over twice as long to remove organic contaminants with argon as with oxygen. Frequently, mixtures of both oxygen and argon are used for plasma cleaning. Table CL-2 compares the various plasma-cleaning system parameters, as well as their reported effect on wire bonds. From these data, it is apparent that a wide range of parameters produces satisfactory cleaning. Parameters, such as an RF power of 100 to 200 W, a gas pressure of 0.5 Torr for either argon, oxygen, or mixtures, with about 10 min of cleaning time have been shown to increase both the bondability and the reliability of wire bonds. For removing thick layers of epoxy bleed or other contaminants, more time or power may be required. If the device is easily damaged, using O_2 plasma with 75 W of RF power for 3 or 4 min may be adequate. Cleaning optimization procedures and schedules were worked out for bonding by Bonham and Plunkett [CL-18] and White [CL-19]. RF powers above 300 W can be detrimental because of excessive heating of the samples and/or by sputtering off the metallization, and can possibly change the electrical characteristics of the devices.

Studies with oxygen plasma (as with UV-ozone) have separated the atomic oxygen, O, and ionized O_2 from the RF plasma (down-stream cleaning [CL-29]), resulting in effective cleaning of such materials as photoresist. Procedures as simple as putting the devices inside screen enclosures (Faraday shield) within the RF plasma will shield sensitive devices from electric fields and prevent radiation damage, as discussed in Appendix CL-1. Various specifically designed RF and microwave down-stream cleaners are available [CL-29]. Unfortunately, no bonding experiments have been performed using these methods. Considering that there is no sputtering and many of the activated atoms will decay along the extended diffusion path, it may be presumed that the cleaning time would be significantly increased over normal plasma cleaning. No downstream cleaning would take place with argon gas, and any silver blackening (see section 4.1.5) could not be removed.

4.1.4 Discussion and Evaluation of Molecular and Solvent Cleaning Methods

Both UV-ozone and plasma cleaning methods improve the bondability as well as the reliability of wire bonds. For ultrasonic and thermosonic bonding, they allow one to use less ultrasonic power and still make a strong weld. This in turn will reduce the incidence of another bond failure (see section 5.1). In addition, strong Au-Al welds have been consistently shown to be more reliable than weak ones. There have been far more published studies using plasma than UV-ozone cleaning for bond quality improve-

TABLE CL-2
VARIOUS REPORTED PLASMA CLEANING PARAMETERS

R.F. Power (Watts)	Gas	Pressure (Torr)	Flow Rate*	Time	Effect On Bonds	Reference
300	O_2	-	300 cc/min	10 min	Reduced corrosion	CL-5
100	O_2	0.5	-	10 min	Increased ball shear strength	CL-7
50	O_2,Ar	0.5		30 min	Cleaned ceramic substrate	CL-17
50-150	O_2,Ar	-	130 cc/min (Various)	2-10 min	O_2 increased reliability Ar removes silver black	CL-18
50-300	O_2	1-2		10 min	Increased bond reliability	CL-19
<300	O_2,N_2,Ar	0.25		300 W, 60 min for Ar	Increased bondability See text	CL-20
100	O_2	-	-	3-5 min	Increased bondability, reliability	CL-22
75-100	Ar	0.2	113 l/min	5,10 min	Increased bondability	CL-23
220	O_2	1	600 cc/min	10-15 min	Increased bondability	CL-26

* Flow rate to achieve a given pressure is dependent upon the volume and characteristics of the plasma cleaner

ment, although both methods are routinely used in production and often merely mentioned in publications without sufficient details to quote.

The first major usage of molecular cleaning methods for bond improvement was in hybrid production [CL-5,11,18,20]. Here, atmosphere impurities resulting from the long storage of chips combined with a great deal of handling and processing (e.g., from epoxy die attach outgassing) led to high wire-bond and corrosion failure rates which made cleaning a necessity. These cleaning methods are also currently used for relatively low-volume expensive VLSI and other IC packaging where high reliability is a requirement. VLSI chips are usually plasma-processed and may have colorless fluorocarbon films on their bond pads which can lead to bondability as well as reliability problems. If they are not cleaned, they can require higher ultrasonic energy for strong bonding, which can lead to cratering, or the displacement of the Al under the bond.

Direct comparisons between UV-ozone, plasma, and solvent cleaning are rare. Sowell et al [CL-8] did compare UV-ozone, argon sputtering, and high-vacuum bake out. From their data (Figure CL-1), the first two appear equivalent in the ability to clean a surface of airborne organic contaminants. However, in some cases, they could not effectively clean unknown contamination from "as- received" glass slides with UV-ozone, but argon plasma cleaning was effective and after that, UV-ozone removed airborne organics. Weiner et al [CL-7] performed the only bondability study that directly compared UV-ozone, oxygen plasma, acid, and complex solvent cleaning methods. They intentionally contaminated both gold and aluminum-bond pads with photoresist, and also the outgas products from two different epoxies. Although slight differences were found between the cleaning methods in removing particular contaminants, they were all essentially equal except for the solvent cleaning. In this case, the bond strength remained as low as for the uncleaned samples. Nesheim [CL-26] showed that chlorinated solvents, e.g., trichlorethane (TCA), can leave free chlorine residues on bond pads. Kenison et al [CL-22] compared oxygen plasma with solvent cleaning for cleaning incoming die and for epoxy bleed removal. They verified Weiner's conclusion that plasma cleaning is effective and that solvent cleaning is not helpful in removing general organic contaminants. Iannuzzi [CL-5] compared various solvent, plasma, and water-cleaning combinations, using biased aluminum triple-track corrosion in an 85 °C/85% RH as the indication of aluminum contamination. She concluded that a freon TMS cleaning step, followed by oxygen plasma and then cold deionized water, was the most effective cleaning combination available. This removed both organic and ionic contamination so effectively that open package, biased aluminum triple tracks withstood 12,000 hours of 85 °C/85% RH without failure! Uncleaned samples all failed within the first hour. Thus, if heavy organic and/or ionic contaminants are suspected, then the combination of freon TMS, oxygen plasma, and cold DI water cleaning is recommended.

4.1.5 Problems Encountered in Using Molecular Cleaning Methods

Both UV-ozone and plasma have been shown to be effective in removing organic contamination from bonding pads, although the degree of effectiveness of each method may vary somewhat, depending on the specific contaminant. Therefore, some evaluation must be made to determine the best choice for a specific application. Detailed studies of the removal effectiveness of a wide range of known contaminants have not been made. It should be noted that some contaminants, such as Cl^- and F^-, can become chemically bound and may not be removed by any of these cleaning methods.

UV-ozone may activate electronic "color centers" in many white Al_2O_3 ceramic substrates resulting in a darkening or yellowing of the surface. This coloration may disappear in a few days or weeks, but generally it will stay indefinitely. The coloration is completely harmless, but customers may be concerned about its appearance. If the device can withstand baking for 200 °C for 8 to 16 hours, the coloration can usually be removed. Similar coloration may result from plasma cleaning, but it is less noticeable and may decay rapidly. Commercial pin grid arrays, multichip packages, etc., are usually dark purple or brown and are not further colored by UV-ozone exposure.

Oxygen plasma cleaning will blacken (oxidize) silver metallization and may reduce bondability. However, Bonham [CL-18] found that changing the oxygen to argon near the end of the cleaning process restored the silver to its original color and regained any bondability loss.

It is well known in plasma processing (reactive ion etching) that the walls of the etching chamber can become contaminated with stable polymers after long usage. These polymers may be redeposited on pads during subsequent operations. Plasma cleaners are subject to the same problems. It is therefore essential to occasionally clean the plasma reaction chamber walls.

There have been informal reports that some special CMOS devices may display increased threshold voltages after plasma cleaning. If the device and its package can withstand thermal stress, then a heat treatment of 200 to 300 °C for 20 to 30 minutes will usually restore the threshold voltage. Degradations of bipolar devices have also been reported. A similar heat treatment will restore the device characteristics. These problems can be reduced and possibly avoided by minimizing the RF power and time. Since

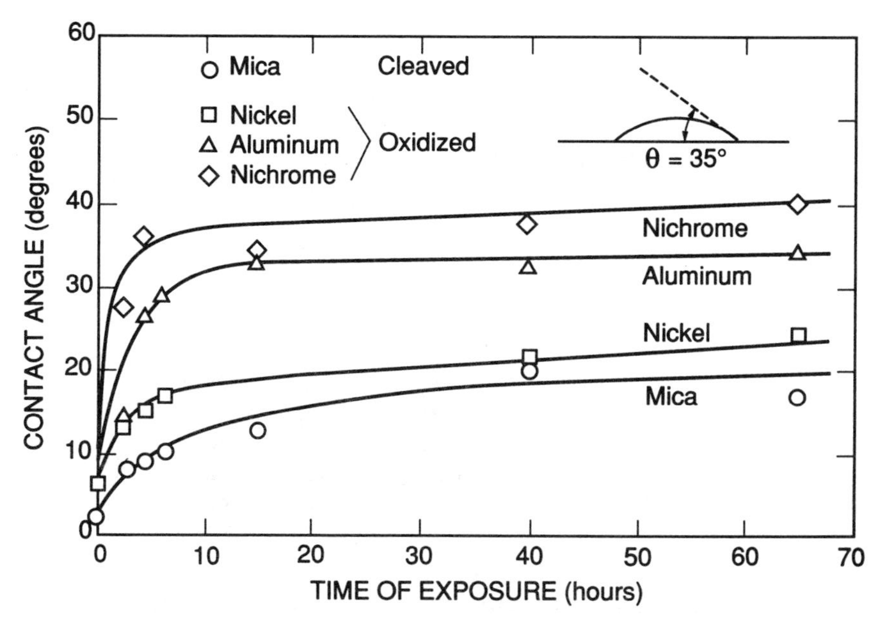

Figure CL-5. A comparison between the recontamination time (in laboratory air) of four surfaces that were cleaned by UV-ozone. Measurements were made by the contact-angle method, illustrated in the upper right portion of the figure. Note that the aluminum (oxidized as it normally is for bond pads) is rapidly recontaminated, but nickel, as used on power device packages, is not (after Vig [CL-12]).

oxygen plasma takes less than half the cleaning time of argon, its use should be encouraged. Problems on both CMOS and bipolar devices are presumed to be the result of the energetic gaseous ions impacting on the device insulators (oxides and nitrides). This can generate electron-hole pairs. The holes may diffuse to active areas, degrading device performance. This phenomenon is called radiation damage. Modern dielectrically-isolated bipolar devices that are not radiation hard are especially affected by such charges. See Appendix CL-1 for a more complete explanation of this effect.

Once a device has been cleaned, it will become recontaminated during storage. Figure CL-1 showed that recontamination of gold surfaces begins within minutes. Vig [CL-12] measured the recontamination time of several materials by the contact angle method as shown in Figure CL-5. It is apparent that different surfaces have different affinities for carbonaceous contaminants from the atmosphere. Of particular interest for bonding is the rapid recontamination of aluminum surfaces. For practical purposes, a period of up to two hours' storage, after cleaning, is acceptable for bonding [CL-7,26]. If stored for longer periods, the device should be recleaned. Storing in nitrogen-filled plastic cabinets may help, but this has not been demonstrated to prolong the clean surface period. The cabinet itself, as well as waffle packs and other plastics inside the enclosure, may outgas organics onto the devices.

4.1.6 Burnishing

Although not generally thought of as a cleaning step, various methods of abrading and scraping surfaces have been used for years to improve the bondability of thick-film metallizations in hybrids.

Thick films may have forms of surface contamination (e.g., glass, metal oxides), as well as pits and voids, that are not removed by molecular cleaning methods. These contaminants or surface defects can reduce both the bondability and the reliability of a bond. Although there is no substitute for a good cleaning step, such as UV-ozone or oxygen plasma, before bonding, burnishing, scouring, and coining thick films to increase bondability have been performed since thick-film technology was developed. Burnishing instruments, such as fiberglass brushes, electric erasers (used by draftsmen) [CL-22], various hard scouring tools, as well as prebonding without wire in the tool (coining) [CL-33], have been used for this purpose. These are controversial procedures. Experiments have been published that support (St. Pierre [CL-31], Panousis [CL-32], Marquis [CL-33]), or criticize (Romenesko [CL-34]) such procedures.

Surface roughness of thick films (not a specific cleaning issue) is considered by all authors to be part of the thick film bondability problem. Coining [CL-33] and electric erasers [CL-22] effectively smooth the surface. Scouring (Figure CL-6) is effective for smoothing, and in addition, it removes all surface contaminants.

The use of various erasers or fiberglass brushes offers the hazard that particles may be left on the surface or imbedded in the thick film and not be removed by other cleaning steps. These then pose a bondability problem on their own. Erasers, in addition to unspecified abrasive particles, can leave rubber particles containing sulfur which may pose a reliability problem if not removed. One would like to think that the use of various erasers and brushes would have diminished. However, these procedures are still regularly found (or thought) to be necessary for bonding to thick films.

If thick films present bondability problems related to surface irregularities or oxide and glass contaminants, then scouring (Figure CL-6) or coining by a bonder, with ultrasonic power turned on, are the best procedures.

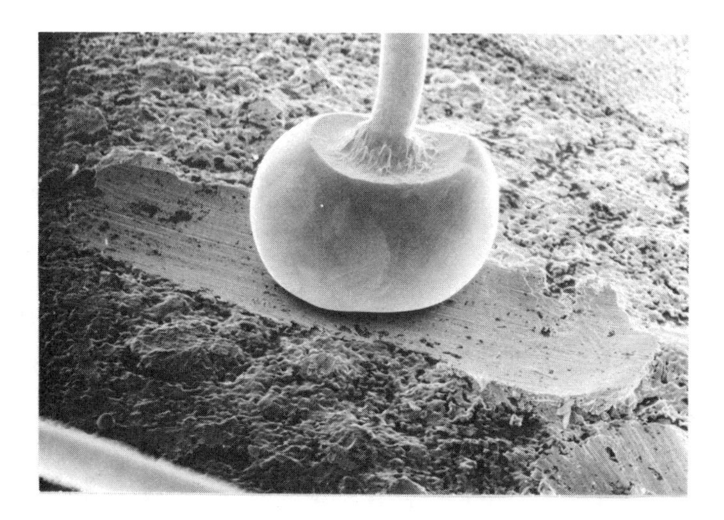

Figure CL-6. A SEM photograph of a 25-μm (1-mil) gold ball bond on a thick film that has been scoured to increase bondability.

4.2 SENSITIVITY OF DIFFERENT BONDING TECHNOLOGIES TO SURFACE CONTAMINATION

The previous section discussed cleaning methods to improve both bondability and reliability. Many of those bondability studies were made with thermocompression bonds, which have great sensitivity to surface contaminants. Most bonding personnel would agree that ultrasonic aluminum bonding is less sensitive to surface contamination than is thermocompression bonding. However, direct comparisons between bonding methods are rare and the experiments difficult to design. The present section compares the bondability of several methods of bonding through various measured thicknesses of a specific contaminant, photoresist. Significant differences between the bonding methods are revealed, and these should be considered when selecting a bonding technology. This is particularly important for complex devices that undergo considerable handling such as hybrid microcircuits.

The only formal study comparing different bonding methods was done by Bushmeyer and Holloway [CL-35] and published in a Sandia National Laboratory report. The substrates used in this experiment were alumina with a chromium adhesion layer and 3 μm of vapor-deposited gold. These substrates were contaminated with spun-on Shipley 1350H photoresist diluted with acetone to various concentrations to produce specific thicknesses ranging from 50 to 180 Å, (effective carbon equivalent thickness of

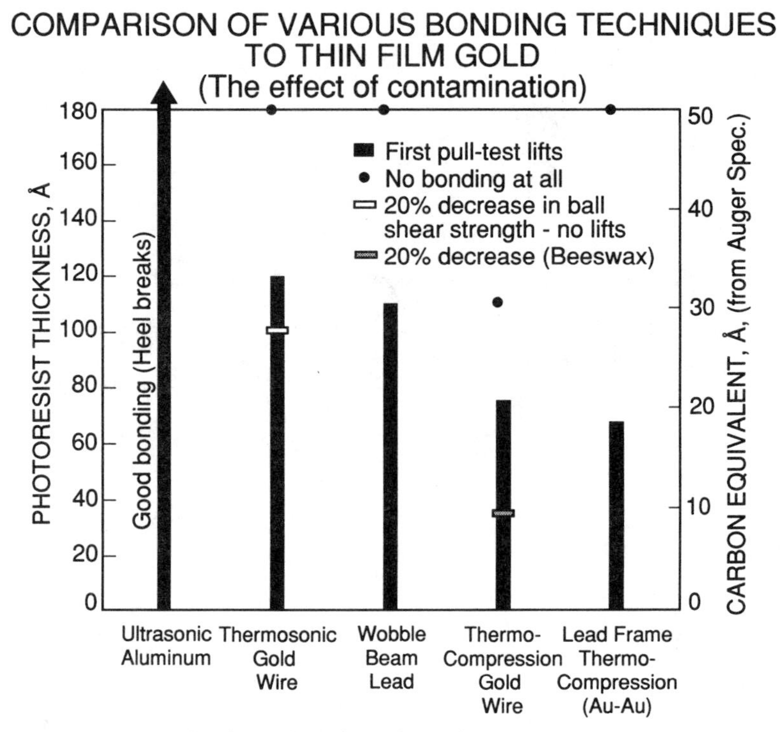

Figure CL-7. Sensitivity of various bonding methods to organic surface contamination. The top of each bar represents the contamination level where the first pull-test lift occurred. The circle indicates the level where no bonding occurred at all.

20 to 60 Å, respectively). The control substrates were UV-ozone cleaned (< 5 Å carbon). All bonding parameters were optimized on the control substrates and then maintained constant for the various contamination experiments. They bonded gold beam leads by compliant and wobble bonding and thermocompression-bonded gold-plated copper lead frames. In addition, they included thermosonic gold ball and ultrasonic aluminum wire bonding.

Data for each contamination level and bonding method consisted of 20 to 40 bonds. Selected results from the study are summarized in Figure (CL-35) drawn from Bushmire's data. The top of each bar indicates the contamination level required to produce the first lifted wire bond in a pull test. Ultrasonic aluminum wire bonding clearly has superior bondability for this organic film contamination. No pull test lifts were observed up to the maximum 180 Å of contamination.

While the results should be valid in the modern electronic device environment, it would be desirable to repeat the experiment using automated bonding machines and a variety of metallizations (e.g., aluminum containing both Si and Cu and various thick films) and with various organic contaminants as well as surface films, such as glass and silicon nitride. There is evidence that during bonding, brittle films break up and are swept into "debris" zones, allowing ultrasonic and thermosonic bonds to be made satisfactorily through relatively thick coatings. Harman [CL-36] found the bondability through 250 Å of CVD oxide unchanged compared with bonding to uncontaminated pads. Clatterbaugh et al [CL-37] found similar results from plasma-thickened Al_2O_3 (up to 200 Å) on aluminum bonding pads.

The above studies were only concerned with bondability. Although strong bonds are more reliable than weak ones, certain contaminants, such as halogens, may affect the reliability of Au-Al bonds and not become evident until later in the device's life.

4.3 APPENDIX CL-1

CIRCUIT DAMAGE CAUSED BY PLASMA CLEANING DURING PACKAGING

Contributed by Peter Roitmam
National Institute of Standards and Technology

Integrated circuits can be electrically damaged by relatively benign plasma processing. This damage can be and is annealed out during the wafer processing sequence by thermal activation. In particular, a near final step in the process is typically a 450 °C anneal to sinter the aluminum metallization. This is sufficient to remove the plasma damage. However, such temperatures are not compatible with modern packaging, and a plasma cleaning step at that time may be harmful.

Radiation damage in semiconductors is usually associated with high-energy particles or photons which penetrate to sensitive junctions and interfaces in the interior of the device. Such high-energy particles or photons can displace atoms in the silicon lattice, resulting in the formation of defects which act as traps or generation centers. If these occur in the base region of a bipolar transistor, the lifetime is reduced and the generation current is increased. This results in failure of bipolar devices. In oxides, the major effect of high-energy particles is to create electron-hole pairs. Holes which drift to the silicon interface are trapped, resulting in an interface charge. If the oxide is in the gate of an MOS transistor, the operating point of the transistor is shifted, resulting in failure [CL-38].

Plasma damage is more indirect. The threshold for ionization of an electron-hole pair in silicon dioxide is 9 eV. Thus, any ion, electron, or ultraviolet photon which has energy greater than 9 eV and collides with an oxide surface can create an electron-hole pair in the oxide. Depending on plasma and surface conditions (secondary emission from the oxide and trapping near the oxide surface region), the oxide surface can charge either positively or negatively. In either case, electrons and/or holes will drift in the field down to the oxide-silicon interfaces. The mobility of electrons is reasonable in silicon dioxide; the mobility of holes is quite low. However, the hole lifetimes are long enough that these carriers can drift a few micrometers. The holes are trapped at the silicon interface creating a positive charge, which can invert the silicon under the oxide. Drift from the top surface of the chip will not normally result in charge under the gate (the classic mode for penetrating radiation damage), but rather in charge under the field oxide. In MOS circuits, inversion of the surface can result in loss of isolation between adjacent transistors, resulting in circuit failure. In modern bipolar circuits, the vertical transistors are oxide-isolated. Thus, charge in the field oxide can short the emitter to the collector. (It is curious how close this mechanism is to that observed by Peck et al in 1963 [CL-39 to 41].) The holes at the oxide silicon interface can be annealed at 300 °C for long times. Below that temperature, they are stable for very long periods [CL-42].

There are a number of other damage modes associated with plasmas such as interface state formation, neutral trap formation, surface damage due to sputtering, etc. These effects can usually be minimized by choice of plasma conditions or else are of minor concern in the field oxide regions [CL-43].

4.4 REFERENCES

CL-1. Thomas, R. W., and Calabrese, D. W., The Identification and Elimination of Human Contamination in the Manufacture of IC'S, 23rd Annual Proc. on Reliability Physics 1985, Orlando, Florida, March 26-28, 1985.

CL-2. Lewis, G. L., and Berg, H. M., Particulates in Assembly: Effect on Device Reliability, 36th Electronics Components Conference, Seattle, Washington, May 5-7, 1986, pp. 100-106.

CL-3. Lewis, R. G., and Wallace, L. A., Toxic Organic Vapors in Indoor Air, ASTM Standardization News, December 1988, pp. 40-44.

CL-4. Lee, W-Y., Eldridge, J. M., and Schwartz, G. C., Reactive Ion Etching Induced Corrosion of Al and Al-Cu Films, J. Appl. Phys. 52(4), April 1981, pp. 2994-2999.

CL-5. Iannuzzi, M., Development and Evaluation of a Preencapsulation Cleaning Process to Improve Reliability of HIC's with Aluminum Metallized Chips, IEEE Trans. on Components, Hybrids, and Manufacturing Technology CHMT-4, No. 4., December 1981, pp. 429-438.

CL-6. Ameen, J. G., Ion Extraction Method Improves Reliability, Proc. of the 32nd IEEE Electronics Components Conference, San Diego, California, 1982, pp. 401-405.

CL-7. Weiner, J. A., Clatterbaugh, G. V., Charles, Jr., H. K., and Romenesko, B. M., Gold Ball Bond Shear Strength - Effects of

Cleaning, Metallization, and Bonding Parameters, Proc. of the 33rd EEE Electronics Components Conference, Orlando, Florida, May 16-18, 1983, pp. 208-220.

CL-8. Sowell, R. R., Cuthrell, R. E., Mattox, D. M., and Bland, R. D., Surface Cleaning by Ultraviolet Radiation, J. Vac. Sci. Technol. 11, No. 1, Jan./Feb. 1974, pp. 474-475.

CL-9. Mittal, K. L., Editor, Surface Contaminator, Genesis, Detection, and Control, Plenum Press, New York, 1979, Vol. 1,2.

CL-10. Jellison, J. L., Effect of Surface Contamination on the Thermocompression Bondability of Gold, IEEE Trans. on Parts, Hybrids, and Packaging PHP-11, No. 3, September 1975, pp. 206-211.

CL-11. Holloway, P. H., and Bushmire, D. W., Detection by Auger Electron Spectroscopy and Removal by Ozonization of Photoresist Residues, Proc. of the 12th Annual International Conference on Reliability Physics, Las Vegas, Nevada, April 1974, pp. 180-186.

CL-12. Vig, J. R., and Le Bus, J. W., UV/Ozone Cleaning of Surfaces, IEEE Trans. on Parts, Hybrids, and Packaging PHP-12, No. 4, December 1976, pp. 365-370.

CL-13. Jellison, J. L., and Wagner, J. A., Role of Surface Contaminants in the Deformation Welding of Gold to Thick and Thin Films, 29th Electronics Components Conference, 1979, pp. 336-345.

CL-14. Clarke, F. K., UV/Ozone Processing: Its Applications in the Hybrid Circuit Industry, Hybrid Circuit Technology, December 1985, p. 42.

CL-15. Zafonte, L., and Chiu, R., UV/Ozone Cleaning for Organics Removal on Silicon Wafers, SPIE 1984 Microlithography Conferences, Santa Clara, California, March 11-16, 1984, Paper No. 470-19.

CL-16. Irving, S. M., A Plasma Oxidation Process for Removing Photoresist Films, Solid State Technology, June 1971.

CL-17. Mead, J. W., Cleaning Techniques for an Al_20_3 Ceramic Wafers, Sandia Report SKND-78-0734.

CL-18. Bonham, H. B., and Plunkett, P. V., Surface Contamination Removal from Solid State Devices by Dry Chemical Processing, published in Surface Contaminator, Genesis, Detection, and Control, K. L. Mittal, Editor, Plenum Press, New York, 1979, Vol. 1,2.

CL-19. White, M. L., Removal of Die Bond Epoxy Bleed Material by Oxygen Plasma, Proc. of the 32nd IEEE Electronics Components Conference, San Diego, California, May 10-12, 1982, pp. 262-265.

CL-20. Graves, J. F., Plasma Processing of Hybrids for Improved Bondability, The International J. Hybrid Microelectronics 6, pp. 147-156 (1983).

CL-21. Graves, J. F., and Gurany, W., Reliability Effects of Fluorine Contamination of Aluminum Bonding Pads on Semiconductor Chips, Proc. of the 32nd IEEE Electronics Components Conference, San Diego, California, May 10-12, 1982, pp. 266-267.

CL-22. Kenison, L. M., Gardner, M. L., and Doering, C. E., Oxygen Plasma Cleaning to Improve Hybrid Wire Bondability, Proc. of the 34th Electronics Components Conference, New Orleans, Louisiana, May 14-16, 1984, pp. 233-238.

CL-23. Buckles, S. L., The Use of Argon Plasma for Cleaning Hybrid Circuits Prior to Wire Bonding, Proc. of the International Symposium on Microelectronics (ISHM), Minneapolis, Minnesota, September 28-30, 1987, pp. 476-479. (This has also been published in other conferences and magazines.)

CL-24. McKee, J. L. J., Toth, W. D., and Fath, P. M., The Characterization and Reliability Prediction of a Thermocompression Wirebonding System, Proc. of the 1986 International Symposium on Microelectronics (ISHM), Atlanta, Georgia, October 6-8, 1986, pp. 259-264.

CL-25. Ebel, G. H., Jeffrey, J. A., and Farrell, J. P., Wirebonding Reliability Techniques and Analysis, IEEE Trans. Components, Hybrids, and Manufacturing Tech. CHMT-5, pp. 441-445 (1982).

CL-26. Nesheim, J. K., The Effects of Ionic and Organic Contamination on Wire Bond Reliability, Proc. of the 1984 International Symposium on Microelectronics, Dallas, Texas, September 17-19, 1984, pp. 70-78.

CL-27. Hansen, R. H., Pascale, J. V., DeBenedictis, T., and Rentzepis, P. M., Effect of Atomic Oxygen on Polymers, J. Polymer Sci. 3, p. 2205 (1965).

CL-28. Khan, M. M., Tarter, T. S., and Fatemi, H., Aluminum Bond Pad Contamination by Thermal Outgassing of Organic Material from Silver-Filled Epoxy Adhesives, IEEE Trans. Components, Hybrids, and Manufacturing Tech. CHMT-10, pp. 586-592 (1987).

CL-29. White, M. L., Detection and Control of Organic Contaminants on Surfaces, Proc. of the 27th Annual Symposium on Frequency Control, Cherry Hill, New Jersey, 1973, pp. 79-88.

CL-30. Cook, J. M., Downstream Plasma Stripping, Solid State Technology 10, pp. 147-151 (April 1987).

CL-31. St. Pierre, R. L., and Riemer, D. E., The Dirty Thick Film Gold Conductor and Its Effect on Bondability, Proc. of the 1976 IEEE Electronic Components Conference, San Francisco, California, April 26-28, 1976, pp. 98-102.

CL-32. Panousis, N. T., and Kershner, R. C., Thermocompression Bondability of Thick Film Gold, A Comparison to Thin Film Gold, IEEE Trans. on Components, Hybrids, and Manufacturing Technology CHMT-3, pp. 617-623 (1980).

CL-33. Marquis, E., and Wallace, A., Surface Preparation of Thick-Film Gold for Automatic Thermosonic Gold Wire Bonding, The International Journal for Hybrid Microelectronics 5, pp. 559-561 (1982).

CL-34. Romenesko, B. M., Charles, Jr., H. K., Clatterbaugh, G. V., and Weiner, J. A., Thick Film Bondability: Geometrical and Morphological Influences, Proc. of the 1985 International Symposium on Microelectronics (ISHM), Anaheim, California, November 11-14, 1985, pp. 408-419.

CL-35. Bushmire, D. W., and Holloway, P. H., The Correlation Between Bond Reliability and Solid Phase Bonding Techniques for Contaminated Bonding Surfaces. Sandia Laboratories Report SAND75-0281, Sept. 1975, pp. 1-23.

CL-36. Harman, G. G., unpublished.

CL-37. Clatterbaugh, G. V., Weiner, J. A., and Charles, Jr., H. K., Gold Aluminum Intermetallics: Ball Bond Shear Testing and Thin Film Reaction Couples, Proceedings of the 34th Electronics Components Conference, New Orleans, Louisiana, May 14-16, 1984, pp. 21-30.

CL-38. Messenger, G. C., and Ash, M. S., The Effects of Radiation on Electronic Systems, Van Nostrand Reinhold, New York, 1986.

CL-39. Peck, D. S., Blair, R. R., Brown, W. L., and Smits, F. M., Bell Syst. Tech. J., 42, 95 (1963).

CL-40. Srour, J. R., and McGarrity, J. M., Proc. IEEE, 76, 1443 (1988).

CL-41. Nicollian, E. H., and Brews, J. R., MOS Physics and Technology, John Wiley and Sons, New York, 1982.

CL-42. Snow, E. H., Grove, A. S., and Fitzgerald, D. J., Proc. IEEE, 55, 1168 (1967).

CL-43. Mogab, C. J. in VLSI Technology, S. M. Sze, Ed., McGraw Hill, New York, 1983, pp. 303-344.

CHAPTER 5

MECHANICAL PROBLEMS IN WIRE BONDING

5.1 CRATERING[1]

5.1.1. Introduction

One type of bonding failure that is commonly attributed to "overbonding"[2] appears as damage to the semiconductor, glass, or other layers under the bonding pad. This is often referred to as cratering, because in severe cases a hole is left in the substrate and a divot is attached to the wire (see Figure M-1). However, far more frequently the defects are less severe. They may produce no visible damage but can degrade the device characteristics. The

[1]This section was presented and distributed at the 1988 VLSI and GaAs Packaging Workshop, Santa Clara, California, September 12-14, 1988. It has been partially updated to include the IEEE Electronics Components Conference presentations [C-12,28].

[2]Overbonding is a term applied to bonding machine parameter setups in which one or more of the bonding parameters (force, time, ultrasonic power, and/or temperature) is significantly greater than is required to produce a normal bond. Usually, this results in the bond being overflattened.

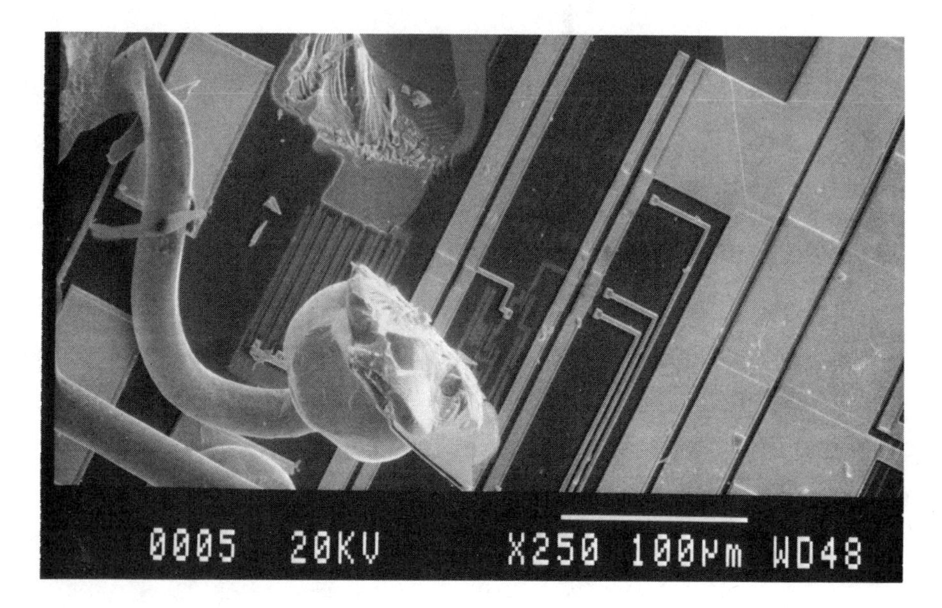

Figure M-1. A SEM photomicrograph of cratering of a bond area with the divot attached to the ball bond. This crater occurred during ball shear testing.

device failure is then incorrectly classed as an electrical rather than a bonding problem. This entire range of damage is referred to as cratering.

Although this failure mechanism is attributed to "overbonding," there are many materials and equipment problems that can enhance the frequency of its occurrence. A study of each case usually reveals a ***synergistic*** relationship between the materials, the bonder setup, and/or later stresses such as plastic encapsulation cure or surface-mount thermal shock. To troubleshoot this problem, one must understand the fracture and deformation of the semiconductor material, possible metallurgical interactions, as well as the wire and bonder characteristics.

There have been two published papers exclusively devoted to studying the causes of cratering during aluminum ultrasonic (US) bonding and three concerning thermosonic (TS) ball bonding. However, there are many papers that have discussed cratering in some other context, such as developing bonding machine setup schedules or using bonding wires of unconventional materials (e.g., copper or silver). A compilation of factors that contribute to cratering is shown in Table M-1. These factors occur in ultrasonic aluminum or gold ***wedge*** bonding as well as in gold, copper, and silver ***ball bonding***. Cratering usually occurs in only a small percent of the

TABLE M-1
CAUSES OF CRATERING

(1) *Materials (metallurgy)*

Wire hardness [M-1, 4, 7, 9, 11]—the harder the wire, the more likely cratering.
Bond pad thickness [M-1, 4, 6]—the thinner the pad, the more likely cratering.
Bond pad hardness [M-9]—ambiguous, but hard pad-interfacial layers (Ti, W, [M-2, M-6]) help.
Stress due to Au-Al intermetallic phases [M-11, 12] can occur after thermal treatment (post mold curing).
Bond pad silicon precipitates [M-2, 5, 15] cited as cracking under glass layers.
Location of pad on device (probably a corrosion-susceptible location difficult to bond).

(2) *Materials (substrate-device)*

GaAs craters easily-yield strength and fracture toughness a factor of 2 lower than silicon [M-6, 18–21].
Bonding to pads over polysilicon [M-8]—polysilicon can separate from underlayer.

(3) *Bonder and Its Set up Characteristics*

Shape or characteristics of ultrasonic generator pulse [M-6, 2, 8]—stable, slow rise time is best.
Ultrasonic energy—too high is harmful [M-1, 4, 5, 7, 8, 11, 27], contaminated bond pads require more US energy to bond.
Bonding temperature—low and high are harmful-ambiguous [M-7, 8, 9].
Shape of tool face for aluminum wedge bonding—concave is best [M-2].
Tool-wire pad impact force [M-1, 2, 28]—limit tool bounce, moderate high impact desirable for ball bonding.
Static bonding force—too high and too low are harmful [M-1, 4, 5, 7, 8, 28].
Negative electronic-flame-off for ball bonding—much better than positive flame-off.

(4) *Failure Symptoms*

Marginal cratering (causes leakage between under layers or in active devices).
Gross cratering (divot comes out at bonding, pull, or shear testing).
Thermal cycling often reveals damage.

(5) *Wafer Processes Affecting Cratering*

Bond pad thickness.
Bond pad and interface metal hardness.
Bond pad silicon or Cu, Al, Si intermetallic precipitates.
Bond pads over polysilicon.
Fracture properties of pad underlayers.
Time, temperature, and cooling rates of heat treatments.

bonds even though the bonds are made at the same time and with the same bond parameters. This small percentage complicates studying the problem and requires experiments with large numbers of bonds and statistical treatments of the data in order to gain an understanding of the process.

Most calculations (see Kale [M-1] and Ching [M-2] for instance) involving cratering start with a circle (cylinder or sphere) being pressed against the bonding pad. The radius of curvature establishes the contact area, and one can easily show (using the Hertz theory of contact pressure) that the initial force creates stresses many times the yield strength and/or the fracture strength of the pad and the wire and, in some cases, of the semiconductor. It is implied that this stress is applied to the underlying semiconductor, thus initiating a crater. Actually, the metal yields far below the yield stress of the semiconductor, deforming the wire (ball) and the bond pad metal. As the wire (ball) flattens, the stress drops rapidly to below the metal yield strength. Ultrasonic energy and/or the stage heater softens the metal which further lowers the yield strength, and deformation continues until the bond is mature (Harman [M-3]).

One major problem with the initial contact-area stress interpretation of cratering is that, according to the Hertz model, the cratering should be initiated at the center of the bonded area. Observations of etch pits due to marginal cratering[3] show that the worst damage occurs along the perimeter of the bond, Figure M-2a. This damage has been clearly demonstrated by Winchell [M-4] in steam oxidation studies that revealed ultrasonic bonding-induced stacking faults in silicon as shown in Figure M-2b. Other work with peeled bonds (see Appendix IA-1 and reference [M-3]) has shown that the perimeter is also the area where initial welding takes place. Koyama [M-5] showed that force alone, without ultrasonic energy, would not cause the form of cratering resulting from silicon nodules in the metallization (see below). Thus, the bonding stress that causes stacking faults and other material damage is primarily related to the ultrasonic energy and therefore cannot be modeled by the normal Hertz contact pressure model.[4] That model, furthermore, is inappropriate since it assumes elastic material properties, whereas bonding results in ***major*** plastic deformation of the metals. The fracture toughness, K_c, of the semiconductor (see

[3]To observe, remove the bond and pad with an etch that will not attack the semiconductor, then lightly etch the semiconductor (or glass) with a preferential etch. Observe the etch pit pattern in a microscope with vertical illumination.

[4]This does not rule out the possibility that significant tool bounce on initial contact cannot be a contributing process. Bounce-type impact forces can far exceed steady-state ones and can occur before the metal has time to deform. Yield strength and fracture toughness are strong functions of the loading or strain rate.

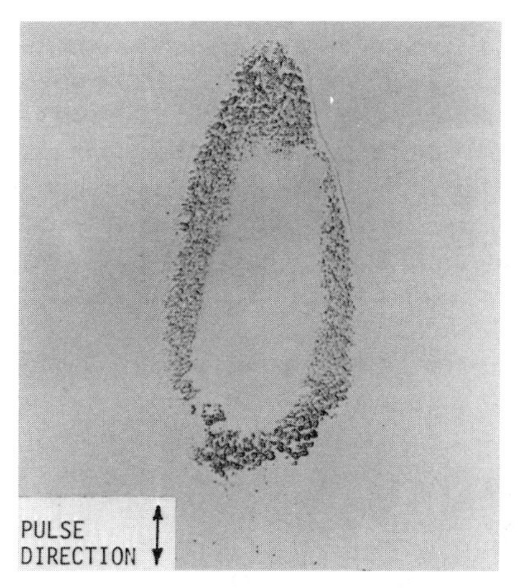

Figure M-2a. A chemical etch-revealed marginal cratering defect pattern on silicon. A 50-μm diameter aluminum wire had been ultrasonically bonded directly to the silicon, then chemically removed and the silicon etched. Magnification 440X (after Winchell [M-4]).

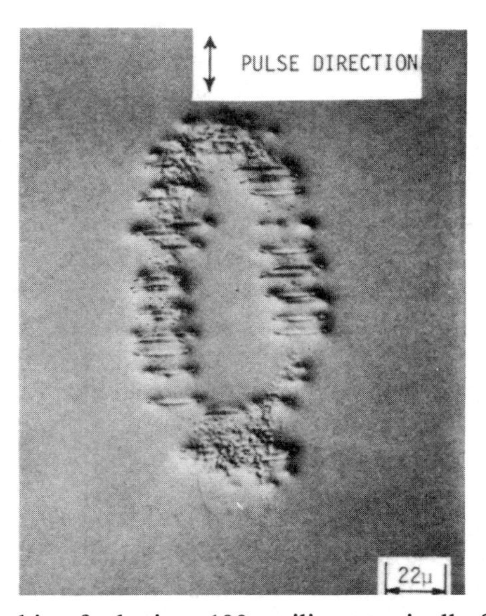

Figure M-2b. Stacking faults in <100> silicon typically form at the bond periphery from forces in the ultrasonic pulsing direction. The bond was made similar to that of Figure M-2a, but steam oxidation was used to reveal stacking faults (after Winchell [M-4]).

Appendix M-1) is the material property closely related to cratering. But, it is not obvious how ultrasonic energy initiates cracks or lowers K_c in a single-crystal, defect-free semiconductor, unless this energy is capable of creating defects as it does in polycrystalline metals. If the stress of an electrical probe or other mechanical damage has produced a microcrack in the position of the bond perimeter, that crack can easily propagate during the ultrasonic welding process. Thermocompression (TC) bonding seldom produces cratering and, therefore, is preferable for use on GaAs devices that are weaker mechanically than Si. If ultrasonic energy is used on GaAs, the bonding process must be carefully optimized and controlled [M-6]. (See section 5.1.9.)

Causes of Cratering

A brief discussion of materials and properties that may not be evident from Table M-1 should help alert the packaging or processing engineer to conditions that may cause cratering.

5.1.2 Bonding Machine Characteristics and Setup Parameters

Excessive ultrasonic energy. The most common cause of cratering results from improper bonding machine parameters, and essentially all papers on the subject implicate this as a contributing cause. Excessive ultrasonic energy has been cited more often than any other bonding parameter as the cause of craters. This is even more apparent when it is considered that cratering is seldom encountered with TC bonding and that this bonding method is the safest process to use on crater-prone materials such as GaAs. In studying the ultrasonic bonding process, Winchell [M-4] found that even though metal mass-flow is equal in all directions, stacking faults in silicon occur perpendicular to the direction of the ultrasonic bonding tool motion (pulse direction), thus verifying that ultrasonic energy is a major cause of the problem and that it is capable of directly introducing defects into single-crystal silicon, although the mechanism is not understood. Koyama [M-5] showed that force and temperature ***without*** ultrasonic energy would not cause the silicon nodule type of cratering.

Modern high-speed automated bonders that can bond up to 10 wires/sec (20 welds/sec) pose an additional cratering hazard. The time available for ultrasonic energy application to each weld has decreased from an average of 50 ms (for older manual bonders) to as little as 10 ms. This generally means that the ultrasonic energy must be increased, or other parameters such as bond force or stage temperature must be increased to compensate. This results in a tighter bond parameter window. However, if the bond

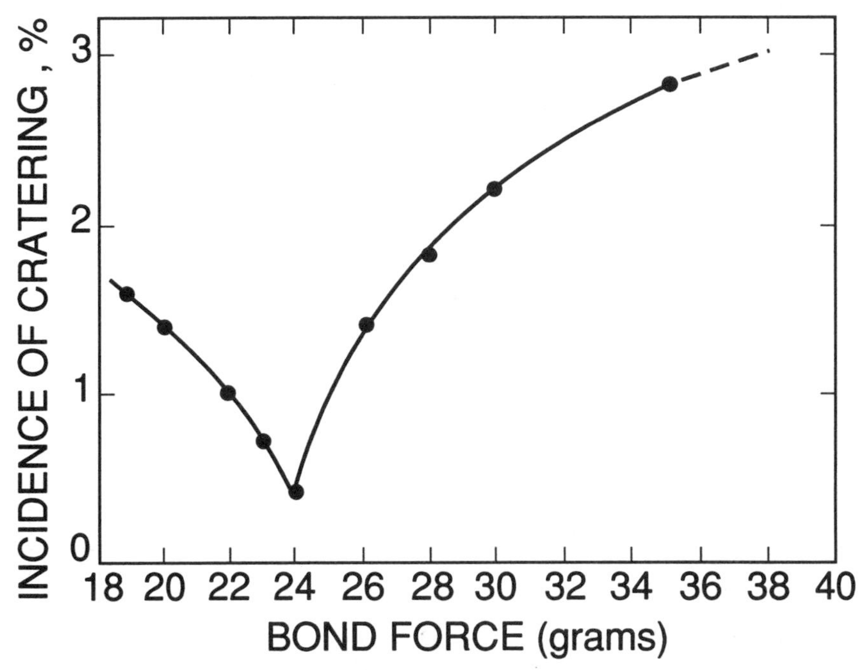

Figure M-3. The incidence of cratering versus bond force for the ultrasonic bonding of 25-μm diameter aluminum 1% Si wire of 15- to 16-gf breaking load. The data were obtained from bonding to various silicon devices (after Kale [M-1]).

time is decreased below 10 ms, increasing ultrasonic energy will be required to produce a reliable weld, and the probability of cratering will increase.

It should be noted that contaminated bond pads require higher ultrasonic energy and/or temperature for making strong bonds. Since modern processing may leave polymer residues on pads, a molecular cleaning method is recommended especially where a cratering problem is encountered (see Chapter 4).

5.1.3 Bonding Force

In general, for wedge bonds, too high *or* too low a static bonding force can result in cratering. This was shown by Kale [M-1] in Figure M-3. The conventional explanation for cratering in wedge bonding with low force is that the bonding tool is not clamped tight enough and it chatters across the top of the wire. This explanation has never been proven. Kale suggested that the optimum force in Figure M-3 resulted in more efficient

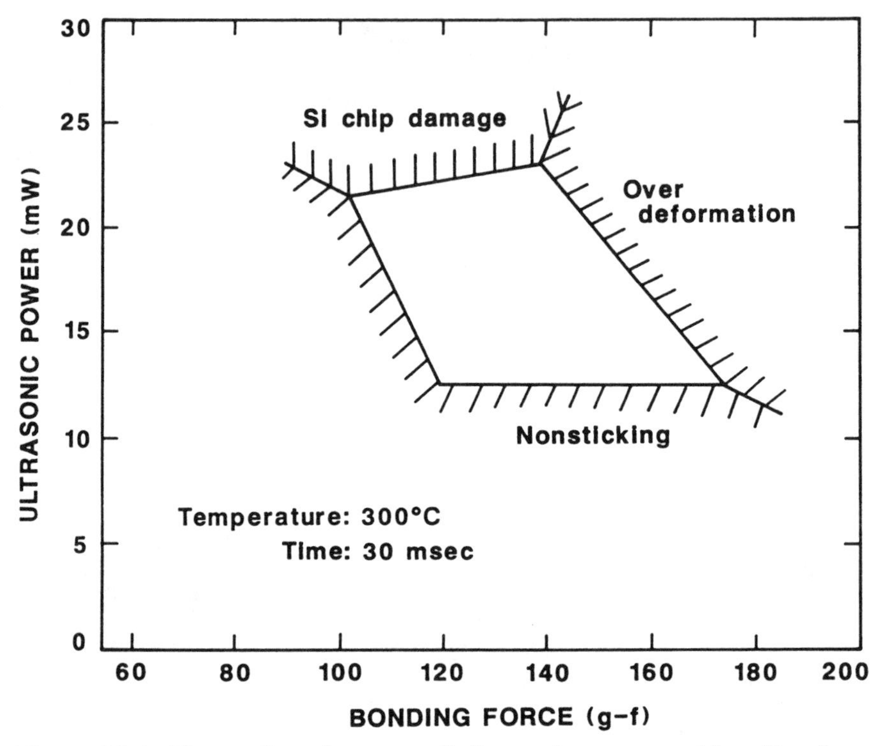

Figure M-4. The preferred ranges of ultrasonic power and bonding force for copper ball bonding to 1.3 μm of Al 1% Si over SiO_2. The initial ball diameter was 62 μm (redrawn from Mori [M-7]).

ultrasonic energy transfer, thus lowering the total energy requirements for the bond. Analysis of data from various sources [M-7,8] indicates that there is a less precise bonding-force effect for cratering during thermosonic (TS) ball bonding. An example of a two-dimensional parameter plot (Figure M-4) for copper ball bonding showing the areas of nonsticking through cratering and overdeformation was given by Mori et al [M-7].

It is difficult to establish clear comparisons between all variables since the bonding parameters must be varied for each experiment [M-4]. If the bond deformation is held constant to make equivalent deformation bonds, then when the force is reduced, the power, time, and/or temperature would normally have to be increased. Most studies do not give enough data to determine the actual parameters causing the damage. Factorial experiments of the type described by Chen [M-8], but specifically designed to encompass the entire range of cratering from crystallographic defects to divoting, should lead to a better understanding of the cratering problem.

Experiments should cover the range from pure thermocompression through pure (no heated stage) ultrasonic ball bonding. Failure analysis of the experiments should be carried out on silicon by etching, possibly by steam oxidation and detailed (optical or electron) microscopic observation. On a more fundamental level, it is desirable to study the mechanism by which ultrasonic energy, during welding, can generate defects (and cracks) in single-crystal semiconductors.

5.1.4 Tool-Wire-Pad Impact Force

Intuitively, one might think that a zero-impact force on the pad would minimize cratering. Low impact is generally used for wedge bonding, especially GaAs, although documentation for this use to minimize cratering has not been published. However, recently, McKenna [M-28] described high-impact bonding as a means of reducing cratering during thermosonic ball bonding. In this case, the capillary and ball rapidly descended to the pad, within 30 ms of ball formation. This left the ball hot (and therefore softer) at touchdown, and it deformed essentially to its mature bond deformation on impact. With the substrate heated to >200 °C, a thermocompression bond would begin to form. Ultrasonic energy (a smaller amount than previously required) applied before or during touchdown matured the bond. The effect of impact force and static bonding force on cratering is shown in Figure M-5. As with Kale's wedge bonding work, too low a static bonding force increases cratering significantly.

At this time, the reason high-impact bonding minimizes cratering is not clear. One may speculate that the combination of elevated ball temperature and impact deforms the ball essentially to mature bond dimensions, initiating a thermocompression bond. The ultrasonic energy applied, starting before touchdown, matures the bond without significant additional deformation. The ultrasonic energy required for this purpose would be less than for a normal bond and, thus, its contribution to cratering would be less also. The soft ball would absorb some of the impact energy as it deformed, applying less force to the underlying material. Note also that the compressive strength of most brittle materials (e.g., glass) is higher under millisecond length impact loads than under static loads.

5.1.5 Causes of Cratering - Materials

Bond Pad Thickness. The bond pad serves as a cushion to protect the underlying material ($Si0_2$, silicon, polysilicon, GaAs, etc.) from damage due to the stresses of bonding. Winchell [M-4] was the first to recognize this. He used an extremely sensitive technique (steam oxidation) to reveal marginal cratering on silicon. He found that "the tendency to crater was

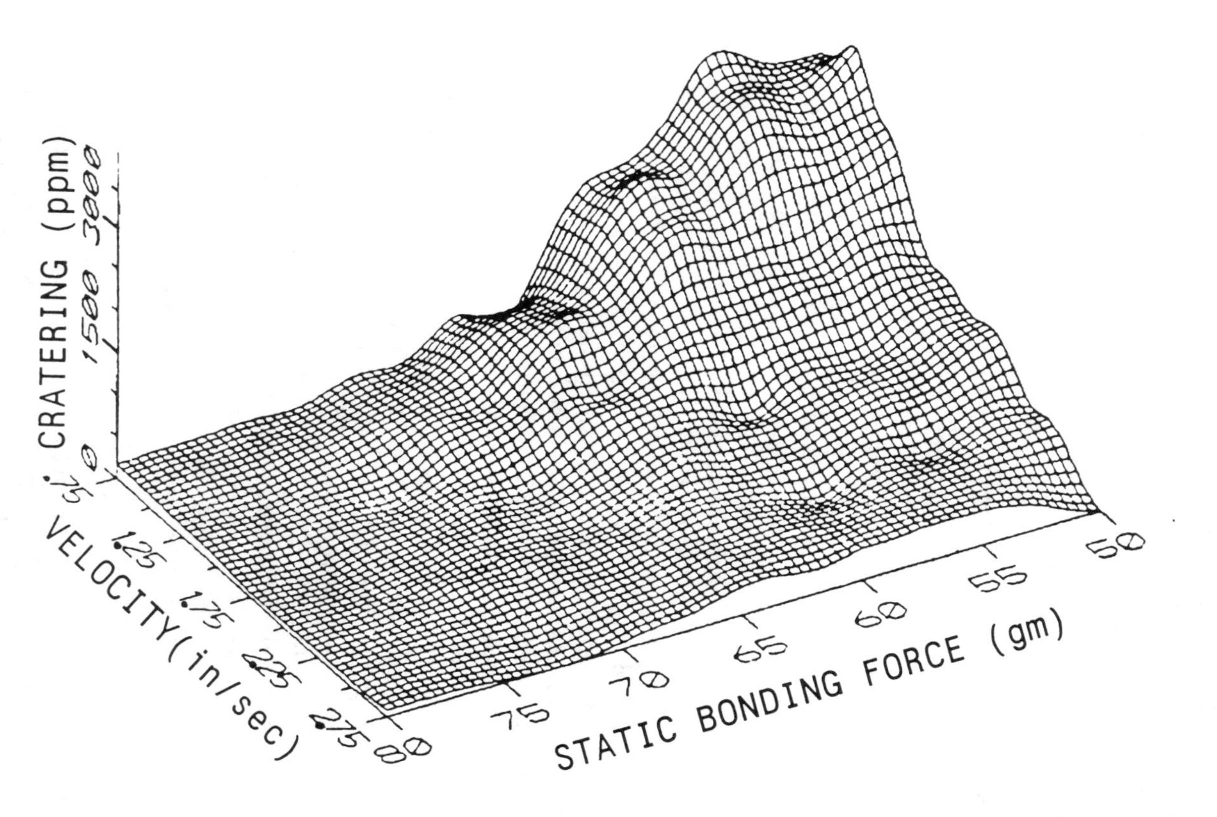

Figure M-5. The effect of impact force and static bonding force on cratering (after McKenna [M-28]).

most prevalent for the thin metallizations." The 0.6-μm metallization thickness represents the transition range in which surface damage to the silicon is still observed. For 1.0- to 3.0-μm metallization thicknesses, the surface damage becomes undetectable when using acceptable bonding machine parameters.

Lycette [M-6] also reported that an increase of the total metallization thickness from 0.8 to 1.2 μm significantly decreased cratering in GaAs devices. Thus, a metallization thickness of 1 μm or greater is desirable to minimize cratering damage in GaAs as well as in Si. Unfortunately, modern VLSI metallization is often thin (≈ 0.6 μm) to facilitate etching of narrow line widths, and this increases the probability of cratering.

Bond Pad Hardness. There is no clear evidence that the bond pad hardness affects the incidence of cratering. One might assume that a softer bonding pad metal would inhibit cratering by absorbing US energy and deforming easily, and that a hard pad should more readily transmit the bonding forces to the substrate. However, as pointed out by Hirota [M-9] and much earlier by Ravi [M-10], the best bonds with the lowest machine bonding parameters are made when the hardness of the wire and the pad are reasonably matched. In fact, the combination of a normal, aluminum layer for bonding over a hard interfacial layer (Ti, W, etc.) appears to be least subject to cratering in both Si and GaAs. Hard copper-doped metallization can be more craterprone because copper oxide or Cu-Al corrosion products on the surface require more ultrasonic energy to bond rather than hardness of the film (see section 2.3.6). The most significant factor may be that the conditions which allow best bond formation also minimize cratering.

Wire Hardness. It has long been known, but not clearly documented, that harder wire can cause silicon craters during aluminum ultrasonic bonding. Kale [M-1] ran cratering experiments with wires of almost equal hardness (as indicated by breaking load) and the data on cratering were ambiguous; nevertheless, presumably based on experience, he listed hardness as a cratering contributor. Winchell [M-4], also using aluminum wire, stated that the frequency of cratering tended to increase with wire hardness.

Recently, there has been an effort to introduce copper and silver wire for ball bonding as a replacement for gold. Since copper and silver balls are significantly harder than gold balls and result in more frequent craters, several investigators have studied the relationship between wire hardness and cratering [M-7,9,11]. (See Figure M-4 as an example of bonding machine setup for copper ball bonding.) Investigators have adopted procedures such as minimizing ultrasonic energy and heating of the substrate and the tool (to soften the ball) to help prevent or minimize the problem. Ching

TABLE M-2
HARDNESS AND SHEAR MODULUS OF WIRE BONDING MATERIALS

Metal	Hardness Value [Material]	Load	Reference	Shear Modulus [M-11]
Gold	40(HV)[ball]	1 gm	M-7	26 GPa
	58-60(HK)[wire] 37-39(HK)[ball]	5 gm	M-26	
	60-90(HV)[bulk]	5 kg	M-24	
Silver	61(HK)[bulk]	15 gm	M-23	30 GPa [bulk]
Aluminum	35-60(HK)[bond pad] 20-50(HV)[bulk]	15 gm 5 kg	M-9 M-24	26 GPa [bulk]
Copper	47-53(HV)[ball]	1 gm	M-7	48 GPa [bulk]
	55(HV)[ball] 77(HV)[wire]	0.5 gm	M-9	
	47-50(HK)[ball] 64-68(HK)[wire]	5 gm	M-26	
	99(HK)[bulk]	15 gm	M-23	

[M-2] found that silicon nodule-induced cratering (section 5.1.7) was reduced if the time for ball formation (EFO spark) to pad touchdown was reduced to less than 30 ms. This left the ball hot (and therefore softer) at touchdown. Such a rapid capillary descent would result in significant impact deformation of the ball and therefore would require less ultrasonic energy for welding. Thus, this softness plus less US energy [M-5] then could produce the observed decrease in silicon nodule (or any) cratering.

Table M-2 compares published hardness measurements made on various bonding wires and metallization. Some measurements are made directly on the ball. It is very difficult to interpolate between the results of different hardness tester indenters and loads, so the most meaningful information is obtained from data comparisons made by the same investigator. The actual role of wire hardness in cratering is not understood, as pointed out by Winchell, because of the synergistic effect of other variables. Harder wires do require more ultrasonic energy to bond,[5] and this energy use

[5]The ultrasonic energy, E, required for bonding is empirically related to the metal hardness by: $E = K(HT)^{3/2}$ where K is a constant, H is the hardness in Vickers, and T = metal thickness. See, for instance, American Welding Society Handbook, V3, 1980. This specific relationship has not been verified for samples with microelectronic dimensions, but a similar relationship is assumed to be valid.

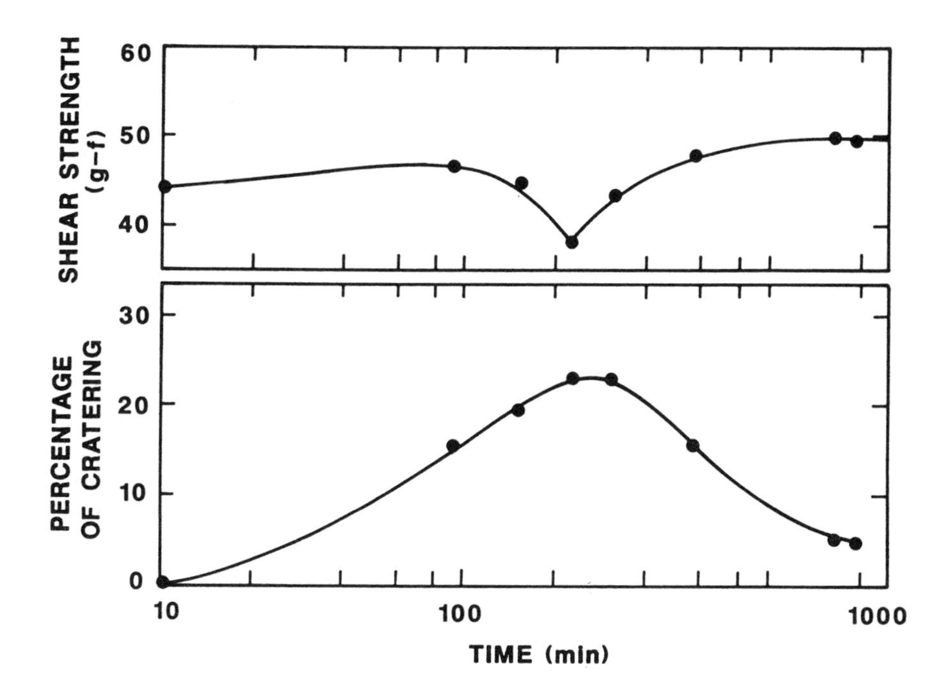

Figure M-6. Percentage of cratering and shear strength for thermally aged (250 °C) gold ball bonds on bonding pads of 1 μm of pure aluminum over silicon (after Clatterbaugh [M-12]).

could be the reason for cratering. Papers on the subject do not give enough information to determine the actual cause. Further work is necessary to fully understand the mechanism.

5.1.6 Intermetallics

Clatterbaugh et al [M-12] studied Au-Al intermetallics primarily by using the ball shear test as a measurement tool. They found that the probability of cratering, as observed in shear testing, increased up to about four hours during a 250 °C bake where ≈20% of the balls cratered the silicon. Continued heat treatment resulted in a decrease in cratering to a level of 4% at 35 hours as shown in Figure M-6. They suggested that the observed Au_2Al resulted in a large-volume increase over that of the original aluminum metallization and thus caused a high stress under the bond, in the order of 90 gf for a typical ball bond. When subjected to the additional stress of a shear test (40-60 gf), the combination of stresses could result in cratering. The authors explained the decrease in cratering after a longer time at the baking temperature as an annealing process that resulted in recrystallization and thereby reduced the stress [M-13]. Also, a different intermetallic compound, with a smaller volume, might decrease the strain. This intermetallic volume-induced strain will be referred to as the Clatterbaugh effect. In a recent (1989) paper, Clatterbaugh and Charles [M-12] reported the results of finite element modeling of the ball shear test with the pad converted to intermetallic. The shear test can apply approximately 3500-7000 kg/cm^2 (50 to 100K psi). Low profile balls result in the lower force, while tall balls may even exceed the higher force figure. (These values may be reduced somewhat by metal yielding.) This force could easily propagate an existing microcrack.

Some interesting consequences of this Au-Al stress mechanism have not been explored. For instance, plastic-encapsulated devices require a post-mold bake of five hours (minimum) at 175 °C, and a calculation indicates that approximately 13 hours will convert 1 μm of aluminum to an intermetallic compound. However, for various reasons, such as a stabilization bake or an overnight work-shift change, the baking time can be extended in some organizations to 16 or more hours. A mid-range of this time at temperature is roughly equivalent to the 10% cratering time-temperature[6] of Clatterbaugh. Thus, most bond pads in plastic-encapsulated devices are at least partially converted to Au_2Al, leaving the underlying structure highly stressed. In some cases, various overcoatings are applied after bonding as an alpha particle barrier and may be cured at various high temperatures. Any anomaly in bonding, in the metallization or in the plastic molding process, could significantly increase cratering. These various contributing sources of stress applied in different steps by different personnel would be easy to overlook.

Cratering under Au-Al bonds, after thermal exposure, has been generally verified by Kamijo [M-11]. His test method (pull test) and baking temperature (400 °C) were quite different from those of Clatterbaugh [M-12], so it was not possible to make a quantitative comparison, but the cratering effect was clear. Ching and Schroen [M-2] failed to observe this effect, but again their temperature-time combination (125 °C for 25 hours)

[6] Using the Arrhenius equation, $t^{1/2} = 5.2 \times 10^{-4} \exp(-15{,}900/kT)$, t = time, T = temperature (Philofsky [M-14]). Note this is the average diffusion of all phases. Others have measured much shorter times (≈1-2 hours) for one or another intermetallic to diffuse 1 μm into aluminum at comparable temperatures.

and/or use of the pull test is not directly comparable. Harman has also observed bond cratering during shear testing after a temperature soak.

Kamijo also found the Clatterbaugh effect with silver wire bonded to aluminum. However, there have been no reports of this effect for copper ball bonding, although all three metals (Cu, Au, and Ag) form numerous intermetallic compounds with aluminum. Unfortunately, studies of strain such as those of Takei [M-13] for Au-Al compounds have not been made for Cu, Ag, or other potential electronic interconnection bonding metals. Excess stress from improperly deposited bumps or metal diffusion barriers (i.e., nickel) in TAB (Tape Automated Bonding) has long been known to result in cratering when the TAB bumps are sheared. Stress from tin-copper intermetallics under solder bumps in controlled collapse flip-chip devices have led to underlying SiO_2 or glass fracture. These are other examples of the cooperative effect of different stresses in producing cratering.

Much more work is needed to clarify the Clatterbaugh effect. Some investigators have not observed (or reported) this effect. Variables such as the aluminum film and/or the SiO_2 thickness, dopants including silicon and copper, and different temperature bakes (which result in different intermetallics, different diffusion rate-constants, and possibly different strain reduction characteristics) may all affect the process. For completeness, the work should also include different bonding wires, such as copper and silver, and the effect of bonding machine variables, such as ultrasonic energy, temperature, and time. Identification of the specific intermetallic(s) causing the problem would be necessary for complete understanding.

5.1.7 Silicon Nodule-Induced Cratering

One percent silicon is often added to aluminum metallization in order to prevent back-diffusion of silicon from shallow junctions into that metal. Such back-diffusion could damage the electrical properties of the device. Recently, Koch [M-15] has reported that micrometer-size silicon nodules in aluminum bond pads form and can act as stress raisers and crack the underlying glass during thermosonic gold ball bonding. Later, reliability testing revealed electrical leakage between the pad and silicon. Various aspects of the plastic molding process were shown to add stress to the bond pad. Corrective action to minimize cratering included bonding at higher temperature (250 °C), lower ultrasonic power, reducing molding stress, and removing a fracture-prone phosphorus glass layer from under the pad. Ching [M-2] and Koyama [M-5] have generally confirmed the silicon nodule bond-cratering effect. Ching solved this problem by changing the structure under the pad (using a hard Ti-W submetallization), by modifying the bonding schedules and by using rapid ball touchdown. His experiments apparently did not include plastic encapsulation or application of any other shear forces. Koyama experienced the silicon nodule-cratering effect during surface mount soldering of plastic-encapsulated devices. He also found that ultrasonic energy, rather than force and temperature alone, was essential to cause cratering. This failure mode was enhanced if the plastic package had absorbed water. The thermally expanding "wet" plastic applied shear forces to the ball bonds during the rapid heating, resulting in cratering. These devices may have included added intermetallic stresses from the Clatterbaugh effect, but this was not investigated.

Discovery of the silicon nodule-cratering mechanism raises questions as to why silicon nodules as large as the bond pad thickness should exist. Bonding wire manufacturers have long controlled the grain size of nodules in similar 1% silicon-doped aluminum wire by proper heat treatment (rapid cooling from 350 to 100 °C). It appears that the metallurgy of aluminum 1% silicon is generally overlooked by integrated circuit manufacturers. After metal deposition and patterning, the films are annealed (sintered) at temperatures >400 °C for up to 30 minutes. They are cooled rapidly ***or slowly*** without consideration of the silicon grain size. Various other processes may require high-temperature wafer heating, such as for curing polyimide multilayers. Again, there is no effort to assure a rapid cooling rate to prevent large silicon particles from growing. Many processing documents prescribe slow cooling after such heat steps to prevent strains that may lead to wafer bowing. Some compromises must be made to minimize silicon nodule growth. In addition, other aspects of device reliability may be improved by this compromise. Large silicon nodules will eventually represent most of the silicon which was originally added to the Al to prevent back diffusion from narrow junctions. After nodule formation, such back diffusion could then take place and damage the device.

Several papers, for instance, Pramanik [M-16] and Umemura [M-17], have considered thermal effects on silicon particles in metallization, so the information to understand the problem is available. In studying cratering, investigators should be aware that Au-Al intermetallic stresses on the bond pad (Clatterbaugh effect) might result from the curing of alpha particle barriers, various glob tops, curing of plastic encapsulants, stabilization bakes, etc. Then, these induced thermal stresses combined with other stresses, i.e., plastic package shear forces, silicon nodules, probe damage, and microdefects (marginal cratering due to poor bonding machine setup) can result in cratering where no single effect alone would cause a problem. The use of several percent of copper in aluminum can be expected to form hard copper-silicon-aluminum intermetallics under various circumstances and can be an additional source of crater-forming nodules.

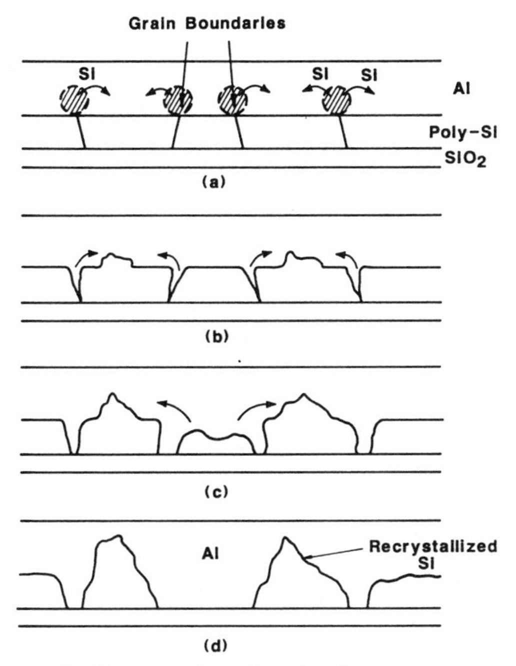

Figure M-7. Schematic diagram showing the time sequence of the restructuring of an undoped polysilicon layer when heated in contact with an aluminum film. The shaded areas in (a) indicate enhanced Si dissolution above the grain boundaries rather than over the grains. The arrows in (a), (b), and (c) indicate the direction of Si transport in Al; (d) final restructured Si. Similar effects occur using Al 1% Si metallization. The recrystallization effect is inversely proportional to the doping level of the polysilicon (after Pramanik [M-16]).

5.1.8 Cratering Over Polysilicon

A high susceptibility for bond cratering under aluminum pads located over polysilicon has been privately reported numerous times and published at least once [M-8]. The cause of this problem is not clear. Obviously, poor polysilicon adhesion to the underlayer, because of processing problems, is possible. Polysilicon also contains more stacking faults, dislocations, and other defects than single crystals, and one can assume that the ultrasonic energy will interact with them and weaken the structure similar to the manner in which it softens metals [M-3]. We propose an additional possible cause of the problem based on studies of aluminum metallization effects on polysilicon by Pramanik [M-16]; see Figure M-7. When aluminum, 1% silicon metallization is sintered (e.g., heated up to 450 °C and up to 60 minutes) in contact with polysilicon, the metal absorbs silicon from the grain boundaries and can result in relocation of the polysilicon.

In extreme cases, this results in large isolated polysilicon grains which may have lower adhesion to the single-crystal substrate or to the bond pad. The regrowth with added aluminum will contain more defects than the original material and, thus, ultrasonic energy could very well lower the fracture toughness of the polysilicon. There are many undetermined variables in this cratering process. The regrowth effect decreases with increased silicon impurity doping levels, and increases with total film heat treatment. There may be additional heat treatments, such as polyimide cure, that can affect the recrystallization. Since these variables are seldom known at the packaging level, the cratering phenomenon would appear to be random, depending on device type or even wafer lot. As with other cratering problems, more studies are required for a full understanding of the processes and necessary controls.

5.1.9 Gallium Arsenide Cratering

Gallium arsenide has long been known to be more susceptible to bond cratering [M-6] and to mechanically induced electrically active defects [M-18] than is silicon. A number of material characteristics have been studied by Forman [M-19], White [M-20], and Vidano [M-21]. Studies of the mechanical properties (where they are linear) indicate that GaAs has approximately a factor of two less strength than silicon. The two major characteristics, hardness and fracture toughness, are the most relevant to cratering and are given in Table M-3. Hardness is a measure of the resistance of a material to deformation. Fracture toughness is a measure of the stress or energy required to propagate a small ***existing*** crack. It is defined in Appendix M-1.

The properties in Table M-3 have been determined by a number of researchers who, in general, were not concerned with the bond cratering problem, but were interested in studying the general mechanical and fracture properties of the material. GaAs is so much weaker than silicon that it is likely to crater in situations that will not affect silicon. It is easy to calculate that the static compressive force on a deformed ball bond (pad contact of 75-μm diameter, applied at the end of bonding by a 50 gf thermocompression bonding load) is approximately 1100 kg/cm^2. This is well over half of the compressive force required to create electrical defects in GaAs and is approaching the brittle fracture stress. Only a small variation in bonding parameters, or the application of ultrasonic energy, can damage the material. Since both the mechanical and fracture properties are weak, craters could originate from either cause. Studies of the crater surfaces are needed to understand the actual cause of the problem. The GaAs strain rate dependence of cratering may be different from Si and should be determined.

The "conventional wisdom" (on cratering) from GaAs chip manufacturers, bonder manufacturers, private communications, as well as one publication [M-29], is that thermocompression bonding is the safest to use. If thermosonic bonding is used, then the highest practical stage heat and minimum ultrasonic energy is recommended. Also, the use of a negative electronic-flame-off for ball formation in thermosonic ball bonding is very helpful to minimize cratering, although the reasons for this are not clear. Ultrasonic wedge bonding is the least desirable and has a very minimal safe bonding window, although it is being successfully used by a number of organizations. In all cases, great care is indicated in bonder setup and monitoring. Lycette et al [M-6] monitored the ultrasonic energy with a capacitor microphone. They found that a slow rise and decay time[7] for the overall ultrasonic system, as well as a thicker metallization with a hard multilayer understructure, resulted in essentially crater-free device production. Figure M-8 is an example of their early cratering failures before the above improvements were made. Fast-rise-time ultrasonic systems have long been blamed for cratering problems due to bonding tool "kick or jumping." However, measurements made with wide band pass capacitor microphones, monitoring the tool tip motion during bonding, have never observed such a kick. Therefore, the explanation for power supply-induced cratering is not clear and should be investigated further.

Although GaAs is far more susceptible to cratering than silicon, the same general procedures that are successful in minimizing cratering on Si are also applicable to GaAs. The use of minimum ultrasonic energy, minimum bonding tool bounce (much more important on GaAs than on Si), and a thick multilayer bond pad structure (thickness >1 μm) are usually successful in reducing the problem. Clean metal requires less ultrasonic energy for bonding; therefore, either UV-ozone or oxygen plasma should be used within 2 hours of bonding (see Chapter 4). Since GaAs has a thermal conductivity about one-third that of Si and an expansion coefficient about twice that of Si, it is important to avoid thermal shock. The use of a stage preheater for thermosonic bonding is desirable as well as the use of a heated capillary, but not the latter alone. Since the mechanical strength and the fracture-energy of GaAs is so low, the possibility of

[7]In many modern computer-controlled automated bonders, the slowed application and removal of ultrasonic power can be programmed after experimentally determining the optimum ramp shape. This can be done regardless of the Q of the ultrasonic power supply.

TABLE M-3
GaAs MECHANICAL PROPERTIES COMPARED TO SILICON

Property	*GaAs - I*	*Si - II* (same Refs and units as I)
Hardness (Vickers HV, 175-g load)	6.9 ± 0.6 GPa [M-19]	11.7 ± 1.5
Hardness* (Knoop HVN, 100-g load)	590 [M-25]	1015
Young's Modulus E	84.8 GPa [M-21]	131
Fracture toughness (energy [indent] +	0.87 j/m^2 [M-20]	2.1
Fracture energy [DCB] + +	1.0 j/m^2 [M-20]	2.1
Compressive force to create elect. defects	1500–2000 kg/cm^2 [M-18]	—

Brittle fracture stress	3.5–15 kg/cm^2 [M-21]	—
Thermal conductivity (300 K)	0.48 W/cm/°C	1.57
Expansion coefficient	5.7×10^{-6}/°C [M-21]	2.3

* Average approx. 20 impressions perpendicular and parallel to the cleavage axes. Taken on same equipment, same operator (John Smith, private communication).

\+ Obtained by indentation after initiation of crack (typical of crater after probe mark or ultrasonic damage). Data from various orientations on the <100> of GaAs and the <111> surface on Si.

\+ + Obtained by double cantilever beam method.

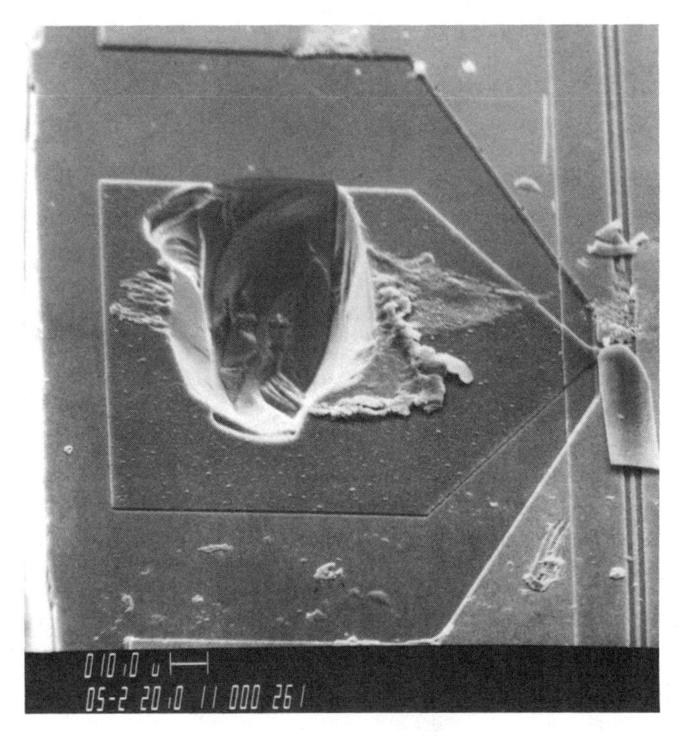

Figure M-8. An example of GaAs cratering under an ultrasonic aluminum wedge bond on a gate pad (after Lycette [M-6]).

electrical test probe damage creating an initial microcrack is much greater than for silicon, and once started, a crack can be readily propagated by ultrasonic energy during bonding.

If the GaAs device has aluminum metallization (rare) and the pad is bonded with gold wire (or the reverse), then the Clatterbaugh effect (after temperature exposure) can apply enough stress (3500 kg/cm^2 calculated for a ball bond) to cause dislocations that degrade the electrical properties of the device [M-18]. Such metal (Au-Al) combinations should be avoided by monometallic bonding, or the lifetime-temperature environment of the device must stay low.

In addition to weak mechanical properties, White [M-20] found that GaAs is susceptible to crack propagation enhancement (a 20% lowered fracture toughness) by environmental influences such as water, acetonitride, heptane, and presumably other common solvents. While this type of crack enhancement is most significant for sawed edges of chips, it could effect probe marks or other subpad damage during cleaning steps within normal assembly line processing. A bond, later placed over that area, could have an increased probability of cratering. It is interesting to note that these solvents either have no effect or actually ***increase*** the fracture toughness of silicon.

5.2 CRACKS IN THE HEELS OF ULTRASONIC WEDGE BONDS

Metallurgical cracks in the heels of aluminum ultrasonic wedge bonds have been another cause of concern to device users [M-30]. An example of such a crack in a 25-μm (1-mil) diameter aluminum wire bond is given in Figure M-9. Cracks can be caused, for example, by using a sharp heeled bonding tool,[8] by operator motion of the micropositioner (if a manual bonder is used) or by bonding machine vibration just before or during bonding tool lift-up from the first bond. However, the most frequent cause is the rapid-tool movement after making the first bond. The tool may rise too high or even progress forward before moving backwards to form the loop. This bends the wire upward from the heel of the bond and then backward, opening up a crack. The heel of the bond is already overworked (weakened) during ultrasonic welding, and one flex forward and backward is often sufficient to form a crack. Such cracks are enhanced if the second bond is lower than the first, typical of reverse bonding, since the wire is bent backwards more than if bonds are on the same level. Also, excessive bond deformation thins and further weakens the heel, which will then crack more easily during loop formation. High loops, desirable for thermal cycling reliability, can lead to greater tool motion and an increased probability of heel cracking.

Device users often feel that heel cracks predispose bonds to early field failure, and this may be the case if the crack is severe. However, many "cracks," when examined at high magnification in a SEM, turn out to be relatively benign tool marks or breaks in the top, amorphous-appearing, surface layer of an ultrasonic bond, as indicated by the arrow in Figure M-10a. The metallurgical defects within this "crack" would be partially, if not entirely, annealed during any subsequent heat treatments, such as glass-ceramic package sealing or a high temperature burn-in. However, the fine inner crack shown in Figure M-10b with its stress-raising inner point may propagate through the wire and cause failure during the device operating life. This crack would be unannealable from the standpoint of thermal-cycle flexure-fatigue life [M-31]. Whether these bonds are annealed or not, there is no reason why otherwise well-made bonds in hermetic packages should fail due to a crack, if the subsequent operating field environment does not include stresses produced by temperature cycling or some other force (such as ultrasonic cleaning) which may flex the wire loops and propagate the crack.

The main purpose of the above discussion is to examine heel cracks objectively. The discussion is not intended to imply a blank-check accep-

[8]A small curvature with a radius approx. 0.3-mil for 1-mil (7.6, 25-μm) diameter wire may help reduce cracking.

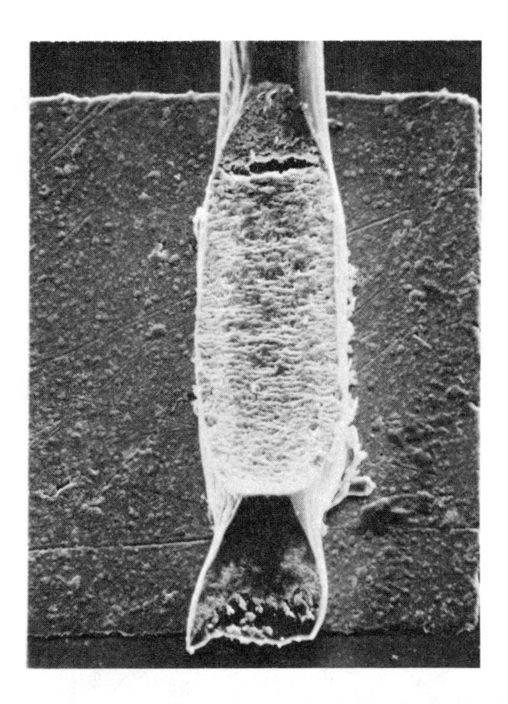

Figure M-9. A SEM photograph of a typical crack in the heel of a 25-μm (1-mil) diameter aluminum ultrasonic wire bond. At this magnification (410X) on this size wire, it is not possible to determine whether the metallurgical defects in the crack are annealable or not.

tance of heel cracks, since some can be so severe that they will significantly degrade the bond pull strength. Any crack of the type shown in Figure M-2b will result in long-term reliability problems under thermal cycle conditions. However, cracks in bonds having small deformation (approximately 1.5 wire diameters) should not significantly reduce the bond pull strength or the device life ***under favorable bond-loop and environmental conditions,*** as is discussed in section 5.4. The existence of such cracks does indicate that some part of the bonding equipment or procedure is not under proper control and corrective action is indicated.

(A)

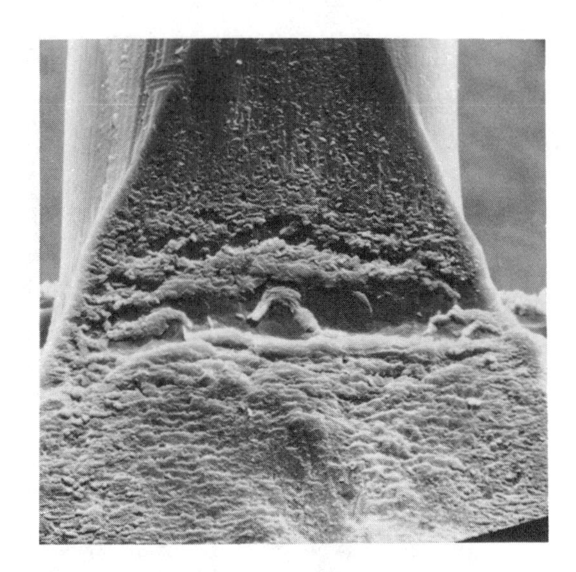

(B)

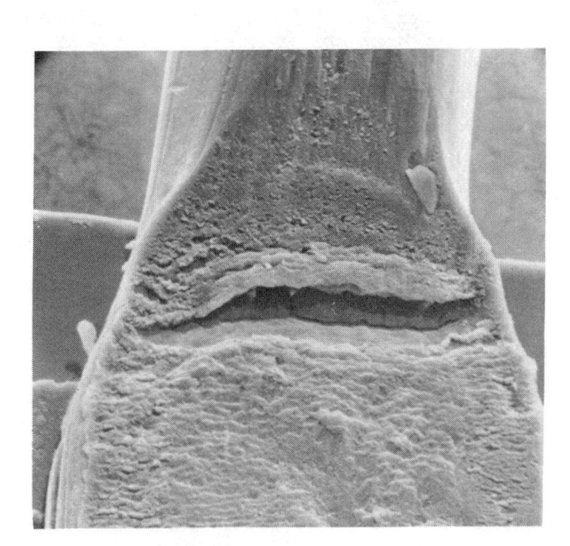

Figure M-10. SEM photographs of cracks in the heels of 50-μm (2-mil) diameter aluminum, 1% silicon, ultrasonic wire bonds. The amorphous-appearing surface layer is indicated by the arrow in (A). No unannealable metallurgical cracks are evident in the exposed crystalline aluminum. The arrow in (B) shows a small unannealable *inner* crack. Both "cracks" resulted from the wire bending forward and then backward during loop formation.

5.3 EFFECT OF ACCELERATION AND VIBRATION

Once the bonding process is finished, the package sealed, and various mechanical, thermal, and electrical screens are completed, the variety of possible mechanical bond-failure modes is reduced considerably. Vibrational- and centrifugal-type forces that occur in the field are seldom severe enough to cause metallurgical fatigue or other bond damage. In general, the package, its leads, or large components of assembled systems will fail before such forces are sufficient to damage the bonds. The ***minimum*** vibrational frequencies that might induce resonance, and thus damage gold or aluminum wire bonds having typical geometries, are approximately 10 and 25 kHz, respectively. The centrifugal forces, in the vertical direction, required to damage well-made bonds of gold or aluminum wire are typically greater than 100,000 g.

5.3.1 Centrifuge Effects on Wire Bonds

Both Schafft [M-32] and Lidbove [M-33] have calculated the effects of centrifuge forces on wire bonds. Their computations were based on the wire assuming the shape of a catenary during the test, and their equations produced similar forces for similar shapes. The equations derived by Schafft are given below. The parameters of the system are the same as those on the bond pull test in section 1.2, Figure T-1.

It may be shown that the tensile forces in the wire, F_{wt} and F_{wd}, at the contact points to the terminal and die, respectively, are in grams-force:

$$F_{wt} = \rho \pi r^2 G (\alpha + h) \qquad (M-1)$$

$$F_{wd} = \rho \pi r^2 G (\alpha + h + H) \qquad (M-2)$$

$$\text{where } \alpha \simeq \frac{d^2}{4h\left(1 + \sqrt{1+(H/h)}\right) + 2H} \quad \text{for } d \gtrsim 2(H+h),$$

and where
ρ = density of the wire (g/cm^3),
r = Radius of the wire (cm),
h = vertical distance between the terminal contact surface and the peak of the wire loop (cm),
H = vertical distance between the terminal and die contact surfaces (cm),
d = horizontal distance between bonds (cm), and
G = centrifugal acceleration (in units of gravity).

Using the approximate value of α will introduce an error of less than 10% in F_{wt} and F_{wd} even for d/(h + H) as small as 2, which is an unusual case. An exact value for α can be obtained from the relation

$$h + H + \alpha + \alpha = \cosh (D/2\alpha),$$

where D/2 is the lateral distance between the bond at the die and the apex of the wire loop. Actually, the greatest uncertainty may result from the estimate of h, which is the distance between the apex of the wire loop and the terminal contact surface after the centrifugal forces have deformed the loop to describe the catenary curve.

A useful graphical representation of equation (M-1) using exact values for α is shown in Figure M-11. Here, the tensile force in the wire adjacent to the bonds of a single-level, 25-μm (1-mil) diameter, gold wire bond subjected to a centrifugal force of 30,000 g's is shown as a function of d for different values of d/h. The figure shows, for example, that for a wire bond with a bond separation of 0.15 cm ($\approx$60 mil) and a loop height of 0.015 cm ($\approx$6.0 mil), the tensile force in the gold wire will be about 0.55 gf. Such a tensile stress could be produced in a pull test if the hook, placed at midspan, were pulled with a force of 0.25 gf. In the case of a gold ball bond, one must take into account the significant mass of the ball—being made up from 0.03- to 0.050-cm (12- to 20-mil) length of the wire. If the force on a ball of 90-μm (3.5-mil) diameter is added vectorially to the wire force in the above example, the total force is increased to $\approx$0.7 gf. In all cases above, if aluminum wire is used, the centrifuge forces will be reduced to 0.14 F_{gold}.

The forces generated in the centrifuge test described above are too small to be useful in testing wire bonds. To obtain an equivalent force to the MIL STD 883 nondestructive pull-test requirement (2.4 gf) on gold wires would require accelerating the device to over 100,000 g, an extreme and impractical value. Consider that a typical gold ball shear force [25-μm (1-mil) diameter wire, 90-μm (3.5-mil) ball diameter] is in the order of 50 gf (tensile force can be 40% higher), and it is obvious that the centrifuge test is useless to assure quality on wire bonds. However, this test can damage gold-wire bonds if the device package is accidentally placed in the centrifuge in the wrong orientation. Poonawala [M-34] investigated the centrifuge problem and found that wire loops can move sideways and short against adjacent wires or terminals, as well as collapse downward and short to the edge of the chip. Wire-loop movement was observed to occur with as little acceleration as 5,000 g and shorts occurred in the 8-10,000 g range. Thus, considerable care must be exercised when hermetic electronic packages are subject to centrifuge testing.

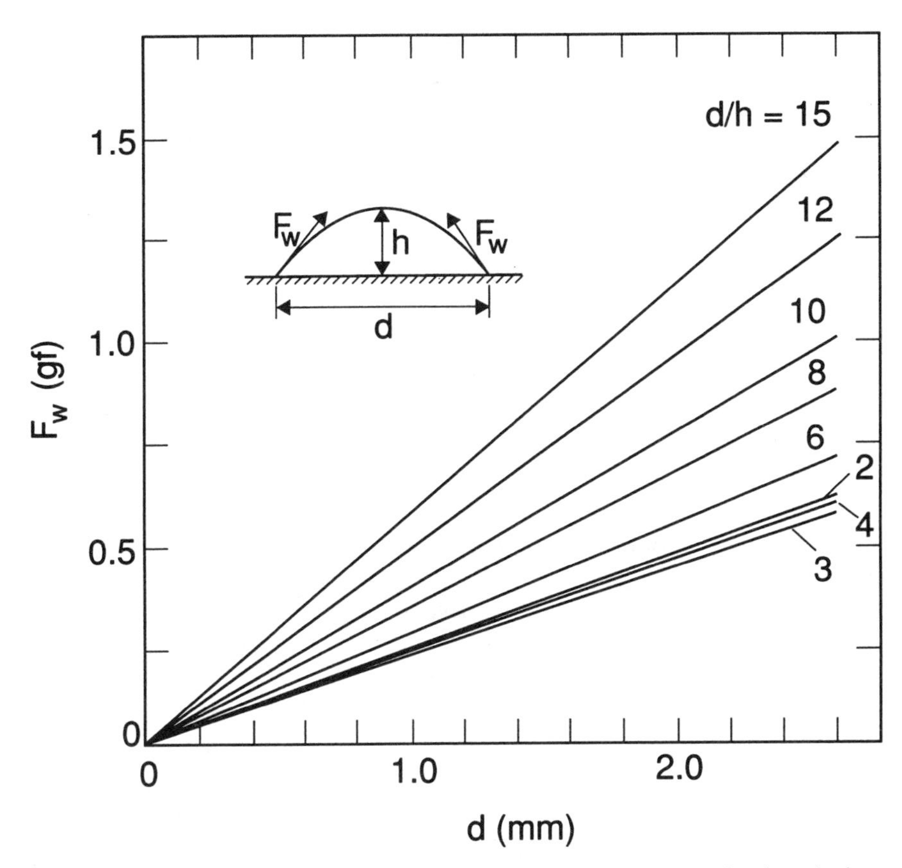

Figure M-11. Tensile force, F_w, in the wire adjacent to the bonds in a single-level, 25-μm (1-mil) diameter, gold wire bond. This resulted from a centrifugal force of 30,000 g's directed perpendicularly away from the bonding surface and is graphed as a function of d for different values of d/h. F_w for a given value of d is a minimum when $d/h \approx 3$. Values for F_w for accelerations other than 30,000 g's may be obtained by multiplying the value for F_w by the ratio of the acceleration of interest to 30,000 g's. Values for F_w for 25-μm (1-mil) diameter aluminum wire bonds may be obtained by multiplying F_w by 0.14 (after Schafft [M-32]).

5.3.2 Effect of Ultrasonic Cleaning on Wire Bonds

There have been several published reports and innumerable private communications on bond degradation resulting from the ultrasonic cleaning of hermetic (open cavity) devices [M-35,36]. Most such reports concern gold-wire bonds, but one [M-37] showed degradation with aluminum bonds.

A wire bond, as with any other wire, has a resonance frequency, and if excited, it will vibrate and may fatigue and break. Figure M-12 is an example of gold-wire bonds that have fatigued during ultrasonic cleaning. The resonant frequencies of wire bonds are high, and about the only sources of such excitations are from ultrasonic cleaners or possibly shock tests. Schafft [M-32] calculated the various vibration modes and the resonant frequencies of wire bonds. He found that several modes are possible; however, the lowest resonant frequency results when the entire loop vibrates side-to-side (lateral mode). Maximum-wire movement would be encountered in this mode, and it would occur at the lowest-excitation frequency. A plot of the calculated resonance frequencies for gold and aluminum wire bonds in various configurations is given in Figure M-13.

The potential danger of bond failure during ultrasonic cleaning occurs because of the many bond configurations encountered on a loaded pc board or even within a given package. A single hybrid circuit can contain bonds made with gold and/or aluminum wire with diameters ranging from 18 μm (0.7 mil) to 38 μm (1.5 mil). The bond-to-bond spacing may range from 0.063 to 0.0038 cm (25 to 150 mil) and the loop heights may vary also.

When many boards are simultaneously cleaned in a large tank, there is the possibility that reflections can cause energy maximum in specific areas. Thus, one would expect that only specific bonds with specific dimensions would be damaged on a given device, and the damage would be maximum in specific areas of the board.

The open package ultrasonic cleaning of hybrids and ICs that will undergo PIND testing for loose particles would be similar to hermetic-package cleaning. The resonant frequency of the bond will shift downward only a few percent due to immersion in the cleaning solvent. However, the liquid would dampen the vibration amplitude depending on its viscosity, but the effect of cavitation could be severe. Informal tests run in various facilities have indicated that the pull strength of some bonds can be degraded or bonds actually broken by this procedure. The effects are very hard to characterize because of the variables involved.

In the past, most ultrasonic cleaners were designed with frequencies in the 20 kHz range and most reported failures, Figures M-11,12 resulting from high-energy industrial cleaners in this frequency range. More recently, ultrasonic cleaner frequencies may be in the range of 20 kHz, 40 kHz, broadband (20 to 100 kHz), or in the hundreds of kilohertz range. Considering the resonant frequencies of bonds from Figure M-12 and that most wire bonds may be within the geometries of curves 2, 3, and 4, it is unlikely that the very high frequency cleaners (>100 kHz) will damage wire bonds,

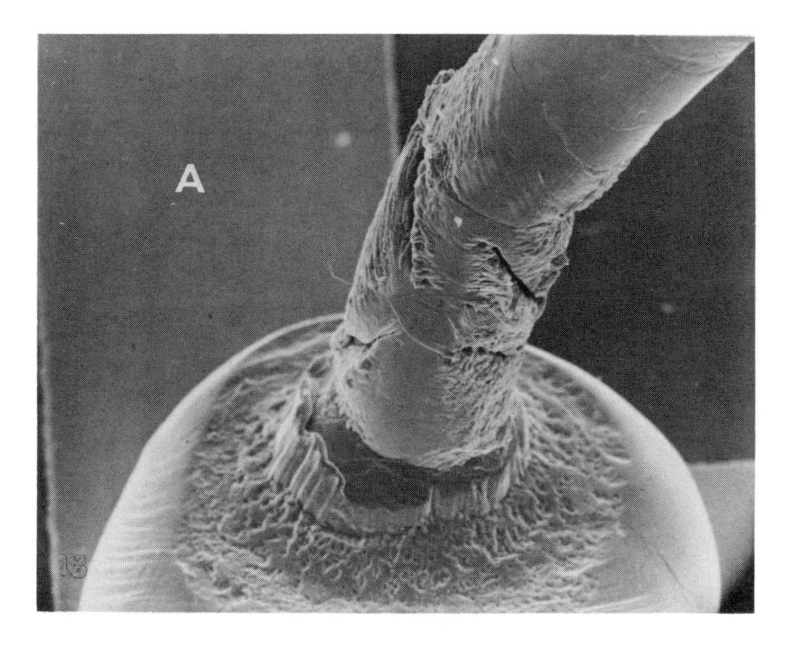

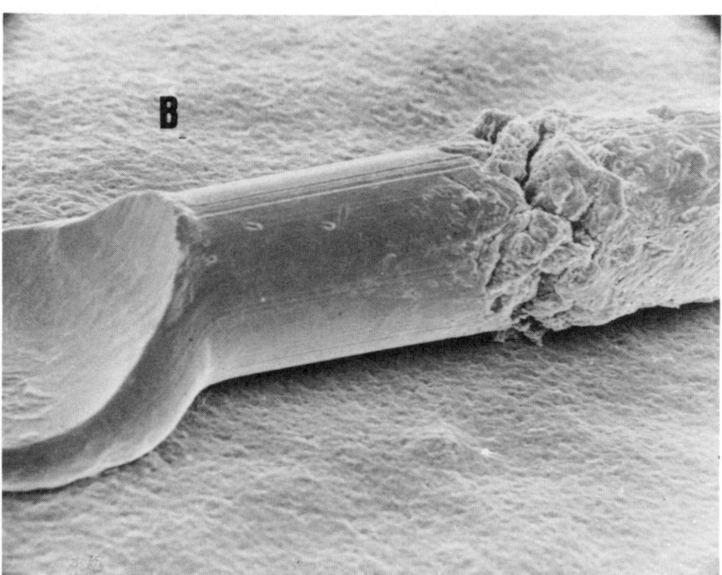

Figure M-12. SEM photographs showing 25-μm (1-mil) diameter ultrasonically fatigued bonds. (A) A thermocompression gold ball bond in a flatpack that had been immersed in an ultrasonic cleaner. (B) A thermocompression wedge bond similarly fatigued (after Harman [M-30]).

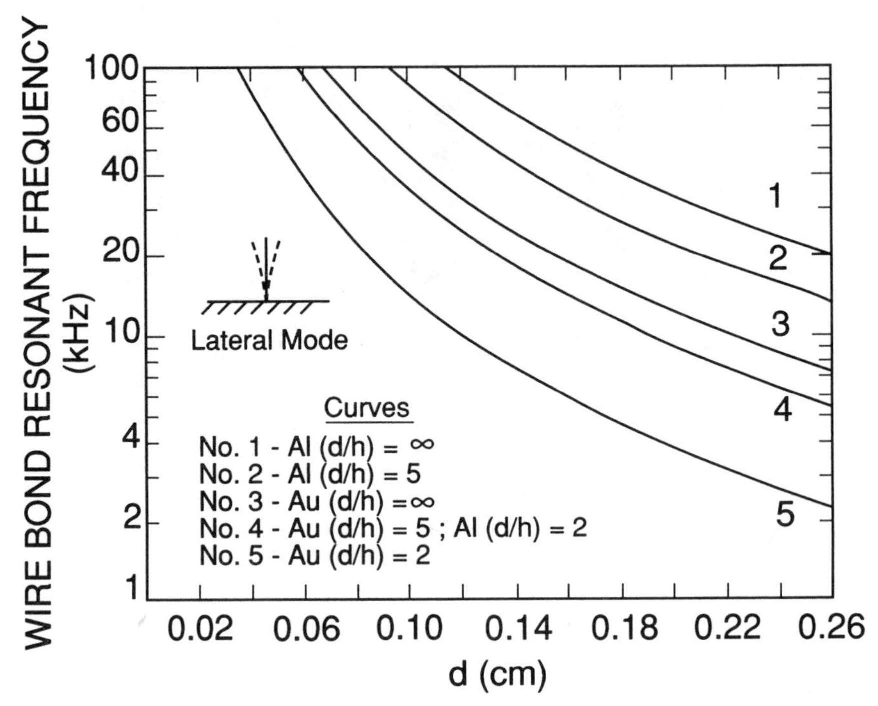

Figure M-13. Dependence of the resonant frequency on bond separation, d, for the lateral mode of vibration for 25-μm (1-mil) diameter gold and aluminum wire bonds with circular-shaped wire spans having the ratios of d/h equal to 2 (semicircular), 5 (intermediate), and infinity (straight wire). The height of the span at its apex is h (after Schafft [M-32]).

although definitive tests to verify this have not been performed.[9] One should be aware that open-package cleaning, at any frequency, is capable of damaging semiconductor devices if cavitation is present.

5.4 EFFECTS OF POWER AND TEMPERATURE CYCLING OF WIRE BONDS

Small-diameter wire-bond failures in open cavity packages due to cyclic temperature changes were first observed by Gaffney [M-38]. There have been extensive studies of such failures [M-39 to 42]. These failures resulted from repeated wire flexing due to the differential coefficient of thermal

[9]At the time that this manuscript is in preparation, an extensive industry-government test of the entire ultrasonic cleaner-induced bond fatigue problem is being conducted at the Navy EMPF Laboratory in China Lake, Ridgecrest, California.

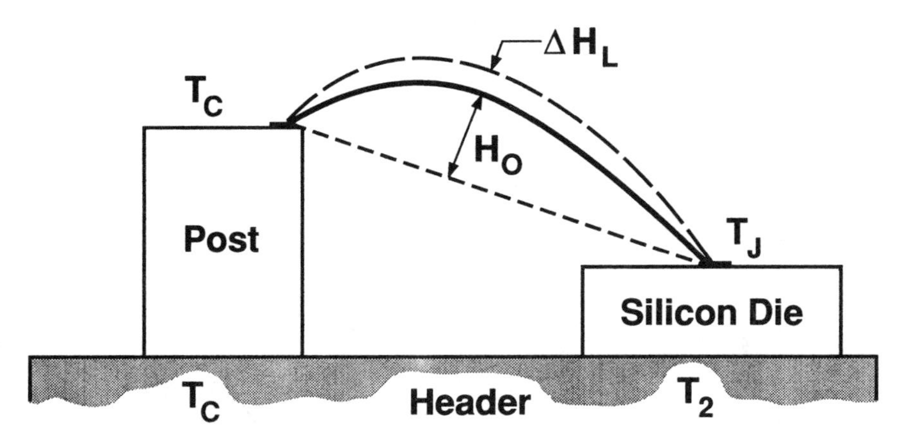

Figure M-14. A schematic representation of wire-bond flexure due to device power cycling. The solid line represents the room temperature position of the bond, and the dashed line represents the high temperature position. The wire may be aluminum or gold, the header may be Kovar, alumina, or other low expansion material. On a first approximation, the expansion of the wire is calculated as the average temperature of T_c and T_J and the header is that of T_c and T_2. The flexing, ΔH_L, is approximately inversely proportional to the ratio of the loop height to the bond-to-bond spacing.

expansion between the aluminum wire and the package as the device heated up and cooled down during power cycling, see Figure M-14. The maximum flexure, and therefore the failure, occurred at the thinned bond heels. The heel of the chip bond experiences the greater temperature excursion and is observed to fail more frequently than the heel of the package bond. Ravi [M-41] experimentally investigated the metallurgical flexure-fatigue of a number of aluminum alloy wires and found that aluminum, 1% magnesium wire was superior to the commonly used aluminum, 1% silicon alloy.

Phillips [M-42] calculated wire-bond geometry effects and found that the flexing increased approximately inversely proportional to the loop height. He recommended that the loop height be approximately 25% of the bond-to-bond spacing to minimize bond flexure of small diameter wire. Figure M-15 shows the results of Phillips' calculation for the flexure of wire bonds with the configuration of Figure M-14. The flexure is approximately 20% greater if the loop is triangular (as it would be after a nondestructive wire-bond pull test). Subsequently, 25-μm (1-mil) diameter aluminum, 1% silicon wire bonds made with high loops survived over 100,000 power cycles even though some bonds had cracked heels [M-42]. Taut

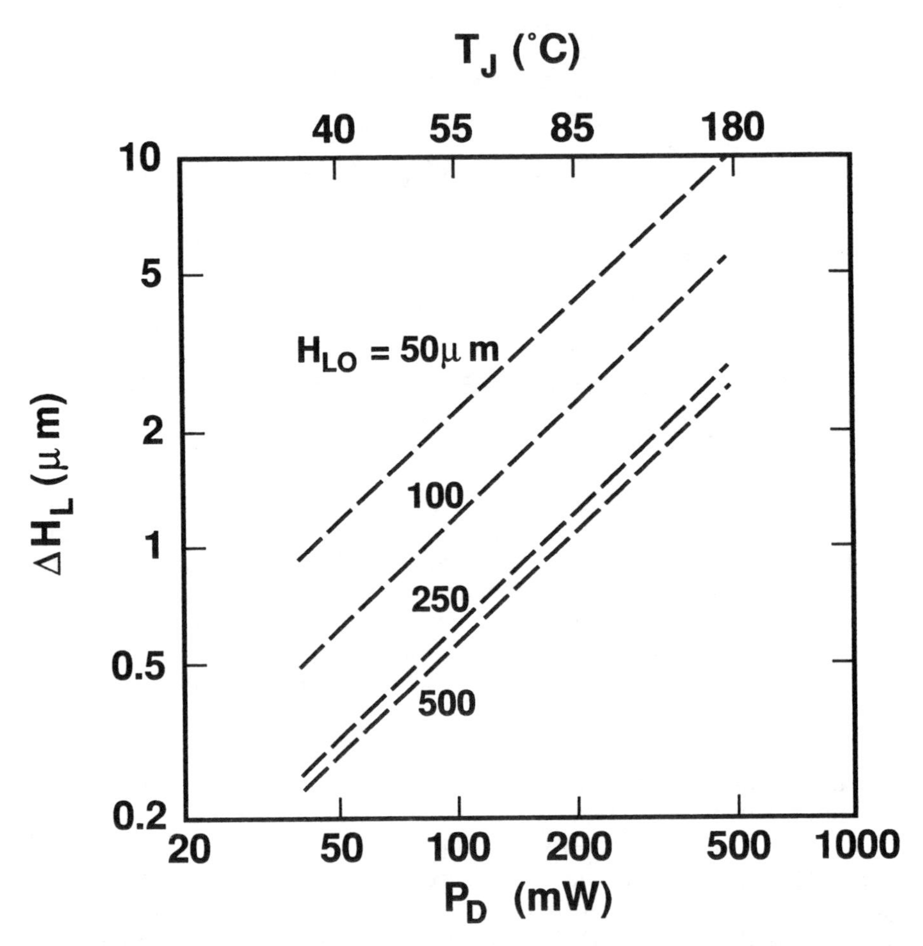

Figure M-15. Wire-bond flexure, ΔH_L, as a function of power dissipation, P_D, for circular arc bond loops with various values of initial loop height, H_{LO}. (The analysis was made for 25-μm (1-mil) diameter post-to-die aluminum wire bonds in a 500-mW, 50-mA silicon transistor in a TO-18 can with post and header of Kovar. The junction temperature, T_J, is also indicated.)

loops resulted in failure after a few hundred to a few thousand power cycles. Villella et al [M-40] have recommended the use of both the magnesium-doped wire and high bond loops in order to achieve maximum reliability. The use of a high loop as protection against flexure fatigue has also been verified with 50-μm (2-mil) in diameter aluminum wire bonds. Figure M-16 shows one typical bond of a high-frequency power device

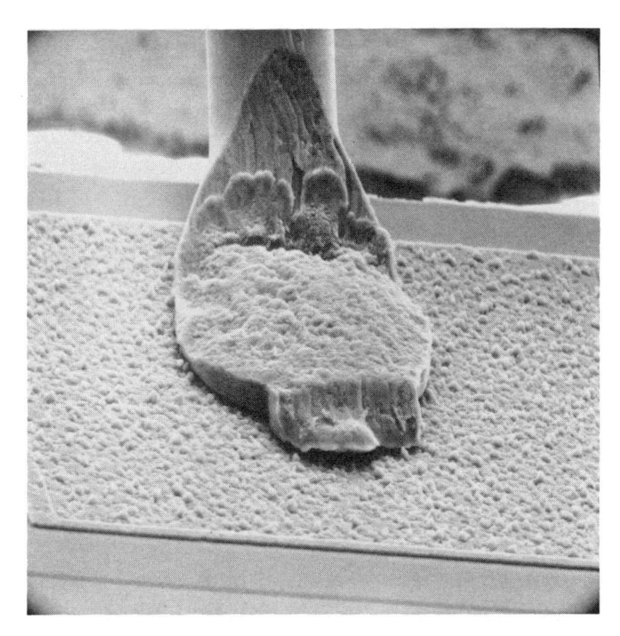

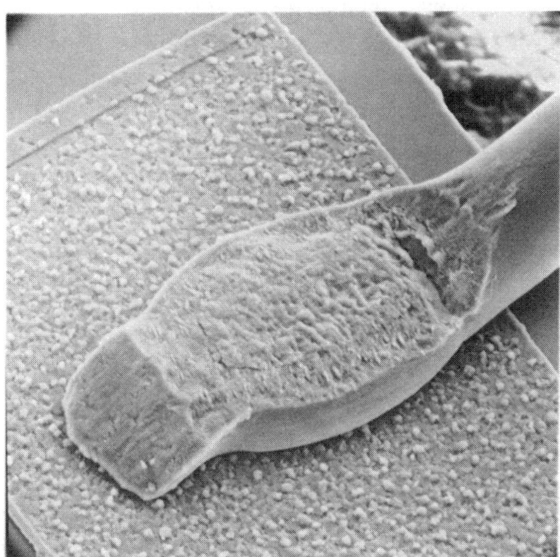

Figure M-16. SEM photographs of power cycled 50 μm (2 mil) diameter aluminum, 1% silicon, ultrasonic wire-bonds with loop heights greater than 25% of the bond-to-bond spacing. The device had undergone 227,627 complete power cycles in which the junction temperature ranged from 38 to 170 °C, and yet no bonds failed. Reconstruction of the bond pad surface resulting from the repeated thermal stress is evident.

containing 47 such bonds with high loops (height greater than 25% of the bond-to-bond distance). This device has undergone 227,627 complete power cycles, without bond failure, in which the junction temperature ranged from 38 to 170 °C. It was characteristic of the production of these bonds that many were made with cracks in their heels (see Figure M-10 for examples of these bonds before cycling), but even so, because of the high loops, no failures were encountered.

Gold-aluminum intermetallic compounds present an additional thermal cycle problem. They are stronger than the pure metals providing they are void-free; however, they are also more brittle (see section 2.1.3). If a bond contains intermetallics, it is far more susceptible to flexure damage than pure gold or aluminum wires alone. An example of a cracked gold wedge-bond which had been cycled only 10 times was given in Figure I-9 in the intermetallics section. In addition to brittleness, the ***growth*** of intermetallic compounds is enhanced by temperature cycling. Thus, it is important to be aware that gold-aluminum couples in devices may fail rapidly during repeated temperature excursions.

While failures in small-diameter wire bonds may be inhibited by high loops, such a solution is only partially helpful for ***large-diameter wires*** where the wire stiffness prevents easy flexing (and thus stress relief) of the overall loop. Large-diameter wire bonds on some power devices may fail after 5,000 to 20,000 power cycles due to metallurgical fatigue. Figure M-17 is a typical example of one of these failures. Note that fatigue occurs at each bend and near the bond heels, which are areas of stress buildup.

Several possible solutions to the large wire fatigue problem may be applicable. The 200-μm (8-mil) diameter wires in this device were 99.99% pure aluminum. According to Riches et al [M-43], alloying additions invariably increase the fatigue strength of alpha solid solutions of aluminum; thus, the better metallurgical system of magnesium-doped wire will help power devices.

The production of smooth loops with no bends should also help. Large cross-sectional-area ribbon-wire together with high loops may improve reliability because ribbon is more flexible in the plane of the bond loop, and thus, may offer the advantages of small wire with high loops.

Temperature cycling (in which the entire device is externally heated and cooled) is more severe than power cycling described above since the entire package reaches the temperature extremes. Even so, the loop height-to-bond length recommendations (25%) are valid and will minimize the effect.

The above discussion of wire-bond flexure fatigue is applicable to bonds in open cavity packages. A variety of different thermal cycle failures have been observed in plastic-encapsulated devices, often where failure occurs above the neck of the gold ball. This failure results from the different thermal

Figure M-17. A SEM photograph of a power-cycled 200-μm (8-mil) diameter 99.99% pure aluminum bonds on a 2N4863 power transistor. The wires fatigued and the device failed at 18,606 power cycles in which the case temperature ranged from 25 to 125 °C (the junction temperature would have reached approximately 180 °C). Metallurgical fatigue occurred at all bonds and at the position of bends in the loops.

coefficient of expansion between the wire, the silicon, and the plastic and is not related to the loop height and wire flexing. The wire is rigidly held within the plastic. Extended fatigue, as in Figures M-17 and M-12, may occur if compliant die coatings are used under the plastic encapsulation [M-44]. For normal single-component encapsulation, the wires often partially "neck down" (ductile fracture) and break. At other times, striations (slip) are seen near the wire break. Ball bonds around the perimeter of the plastic- encapsulated chip may be sheared due to forces exerted by the thermally expanding plastic [M-45] (also see section 5.1.6). In general, there is little that can be done at the wire-bonding production level to prevent such failures. In making low profile ball bonds, the use of low-stress molding compounds and avoidance of thermal shock during soldering to boards are reasonably effective in minimizing the plastic expansion induced bond problems.

5.5 APPENDIX M-1

FRACTURE TOUGHNESS, DEFINITION

Unstable fracture occurs when the stress-intensity at the crack tip, K (or G), reaches a critical value (K_c or G_c). For small crack-tip plastic deformation (plane-strain conditions), the critical stress-intensity factor (K_{IC}) for fracture instability is a material property. The K_{IC} is the maximum stress-field intensity at the tip of a crack that the material can withstand without unstable crack extension occurring. Fracture toughness is a generic term for various measures of the resistance of a material to the extension of a crack. It is frequently represented as K_c, the stress-intensity, or as G_c, the energy release rate for a crack extension [M-22]. The relationship between these variables is given below.

$$U = \frac{\pi a^2 \sigma^2}{E} \text{ (Griffith equation)}$$

U = decrease in elastic energy to propagate a crack due to an existing crack of length a

$$\text{Stress intensity} = K_c = \sigma_c \sqrt{\pi a} \, \frac{gm}{cm^2} \sqrt{cm} \text{ (Fracture toughness)}$$

$$\text{Energy Release Rate} = G_c = \frac{\pi\sigma_c a}{E} = \frac{K_c^2}{E}\frac{gm}{cm^2} \times cm \text{ (Fracture toughness)}$$

σ = Stress $\frac{gm}{cm^2}$; a = crack length (cm); E = Young's modulus

5.6 REFERENCES

M-1. Kale, V. S., Control of Semiconductor Failures Caused by Cratering of Bonding Pads, Proc. of the 1979 International Microelectronics Symposium, Los Angeles, California, November 13-15, 1979, pp. 311-318.

M-2. Ching, T. B., and Schroen, W. H., Bond Pad Structure Reliability, 24th Annual Proc., Reliability Physics, Monterey, California, pp. 64-70 (1988).

M-3. Harman, G. G., and Leedy, K. O., An Experimental Model of the Microelectronic Ultrasonic Wire Bonding Mechanism, 10th Annual Proc., Reliability Physics, Las Vegas, Nevada, pp. 49-56 (1972).

M-4. Winchell, V. H., An Evaluation of Silicon Damage Resulting from Ultrasonic Wire Bonding, 14th Annual Proc., Reliability Physics, Las Vegas, Nevada, pp. 98-107 (1976). Also see Winchell, V. H., and Berg, H. M., Enhancing Ultrasonic Bond Development, IEEE Trans. on Components, Hybrids, and Manufacturing Technology CHMT-1, pp. 211-219, (1978).

M-5. Koyama, H., Shiozaki, H., Okumura, I., Mizugashira, S., Higuchi, H., and Ajiki, T., A New Bond Failure Wire Crater In Surface Mount Device, 26th Annual Proc., Reliability Physics, Monterey, California, pp. 59-63 (1988).

M-6. Lycette, W. H., Knight, E. R., and Hinch, S. W., Thermosonic and Ultrasonic Wire Bonding to GaAs FETs, The International

Journal for Hybrid Microelectronics 5, No. 2, pp. 512-517, November 1982.

M-7. Mori, S., Yoshida, H., and Uchiyama, N., The Development Of New Copper Ball Bonding-Wire, 38th Electronic Components Conference, Los Angeles, California, May 9-11, 1988, pp. 539-545.

M-8. Chen, Y. S., and Fatemi, H., Au Wire Bonding Evaluation By Fractional Factorial Designed Experiment, The International Journal for Hybrid Microelectronics 10, No. 3, 1987, pp. 1-7.

M-9. Hirota, J., Machida, K., Okuda T., Shimotomai M., and Kawanaka R., The Development of Copper Wire Bonding For Plastic Molded Semiconductor Packages, 35th Electronic Components Conference Proc., Washington, DC, May 20-25, 1985, pp. 116-121.

M-10. Ravi, K. V., and White, R., Reliability Improvement in 1-mil Aluminum Wire Bonds for Semiconductors. Final Report (Motorola SPD), NASA Contract NAS8-26636, December 6, 1971.

M-11. Kamijo, A., and Igarashi, H., Silver Wire Ball Bonding And Its Ball/Pad Interface Characteristics, 35th Electronic Components Conference Proceedings, Washington, DC, May 20-25, 1985, pp. 91-97.

M-12. Clatterbaugh, G. V., Weiner, J. A., and Charles, Jr., H. K., Gold-Aluminum Intermetallics: Ball Bond Shear Testing and Thin Film Reaction Couples, IEEE Trans. on Components, Hybrids, and Manufacturing Technology CHMT-7, 1984, pp. 349-356. Also see earlier publications by Charles and Clatterbaugh, International Journal for Hybrid Microelectronics 6, 171-186 (1983). Also, more recently, Clatterbaugh, G. V. and Charles, Jr., H. K., The Effect of High Temperature Intermetallic Growth on Ball Shear Induced Cratering, 39th Proc. IEEE Electronics Components Conference, Houston, Texas, May 21-24, 1989, pp. 428-437.

M-13. Takei, W. J., and Francombe, M. H., Measurement of Diffusion-Induced Strains At Metal Bond Interfaces, Solid State Electronics, Pergamon Press, 1968, Vol. 11, pp. 205-208.

M-14. Philofsky, E., Design Limits When Using Gold-Aluminum Bonds, 9th Annual Proc., Reliability Physics, Las Vegas, Nevada, pp. 11-16 (1971).

M-15. Koch, T., Richling, W., Whitlock, J., and Hall, D., A Bond Failure Mechanism, 24th Annual Proc., Reliability Physics, Anaheim, California, pp. 55-60 (1986).

M-16. Pramanik, D., and Saxena, A. N., VLSI Metallization And Its Alloys, Part 1, Solid State Technology, January 1983, pp. 127-133. Part II, Ibid., March 1983, pp. 131-138.

M-17. Umemura, E., Onoda, H., and Madokoro, S., High Reliable Al-Si Alloy/Si Contacts By Rapid Thermal Sintering, 26th Annual Proc., Reliability Physics, Monterey, California, pp. 230-233 (1988).

M-18. Hasegawa, F., and Ito, H., Degradation of a Gunn Diode by Dislocations Induced During Thermocompression Bonding, Appl. Phys. Lett 21, No. 3, August 1, 1972, pp. 107-108.

M-19. Forman, R. A., Hill, J. R., Bell, M. I., White, G. S., Freiman, S. W., and Ford, W., Strain Patterns in Gallium Arsenide Wafers: Origins and Effects, Defect Recognition and Image Processing in III-V Compounds II, edited by E. R. Weber, Elsevier Science Publishers B. V., Amsterdam, 1987, pp. 63-71.

M-20. White, G. S., Freiman, S. W., Fuller, E. R., Jr., and Baker, T. L., Effects of Crystal Bonding on Brittle Fracture, J. Matls. Sci. 3, 491-497, 1988.

M-21. Vidano, R. P., Paananen, D. W., Miers, T. H., Krause, J., Agricola, K. R., and Hauser, R. L., Mechanical Stress Reliability Factors For Packaging GaAs MMIC and LSIC Components, IEEE Trans. on Components, Hybrids, and Manufacturing Technology CHMT-12, 1987, pp. 612-617.

M-22. Barsom, J. M., and Rolfe, S. T., Fracture and Fatigue Control in Structures, Second Edition (Prentice-Hall, Inc., Englewood Cliffs, New Jersey, 1987).

M-23. Olsen, D., Wright, R., and Berg, H., 13th Annual Proc., Reliability Physics, Las Vegas, Nevada, pp. 80-86 (1975).

M-24. Kashiwabara, M., and Hattori, S., Formation of Al-Au Intermetallic Compounds and Resistance Increase for Ultrasonic Al Wire Bonding, Review of the Electrical Communication Laboratory 17, pp. 1001-1013 (1969).

M-25. Smith, John, NIST, private communication.

M-26. Douglas, P., and Davies, G., The Influence of Electronic Flame-off-Polarity on the Structure of Gold Wire Balls, and An Investigation into the Microstructure and Micro Hardness of Various Ball-Formed Copper Wires, American Fine Wire Corp. Reports, 1988.

M-27. Schulz, G., and Chan, K., A Qualitative Evaluation of Compound Bonds, Proc. of the International Symposium on Microelectronics (ISHM), Seattle, Washington, October 17-19, 1988, pp. 238-245.

M-28. McKenna, R., High Impact Bonding to Improve Reliability of VLSI Die in Plastic Packages, 39th Proc. IEEE Electronics Components Conference, Houston, Texas, May 21-24, 1989, pp. 424-427.

M-29. Riches, S. T., and White, G. L., Wire Bonding to GaAs Electronic Devices, Proc. of the 6th European Microelectronics Conference

(ISHM), Bournemouth, United Kingdom, June 3-5, 1987, pp. 143-151.

M-30. Harman, G. G., Metallurgical Failure Modes of Wire Bonds, 12th Annual Proc., Reliability Physics Symposium, 131-141, Las Vegas, Nevada, April 2-4, 1974.

M-31. Plumbridge, W. J., and Ryder, D. A., The Metallography of Fatigue, Metallurgical Reviews 14-15, p. 129 (1970); also T. H. Alden, and W.A.Backoffen, The Formation of Fatigue Cracks in Aluminum Single Crystals, Acta Metallurgica 9, pp. 352-366 (1961).

M-32. Schafft, H. A., Testing and Fabrication of Wire-Bond Electrical Connections, A Comprehensive Survey, Nat. Bur. Stands. (U.S.), Tech. Note 726 (1972).

M-33. Lidbove, C., Perkins, R. W., and Kokini, K., Microcircuit Package Stress Analysis, Final Technical Report RADC-TR-81-382, Rome Air Development Center, pp. 1-351 (1982).

M-34. Poonawala, M., Evaluation of Gold Wire Bonds in a Cannon-Launched Environment, 33rd Proc. IEEE Electronics Components Conference, Orlando, Florida, May 16-18, 1983, pp. 189-192.

M-35. Ramsey, T. H., Metallurgical Behavior of Gold Wire in Thermal Compression Bonding, Solid State Technology 16, pp. 43-47 (October 1973).

M-36. Harman, G. G., Metallurgical Failure Modes of Wire Bonds, 12th Annual Proc. IEEE Reliability Physics Symposium, Las Vegas, Nevada, April 2-4, 1974, pp. 131-143.

M-37. Microcircuit Manufacturing Control Handbook, Integrated Circuit Engineering Corporation, Scottsdale, Arizona, 1977, pp. P5-J5, 5A.

M-38. Gaffney, J., Internal Lead Fatigue Through Thermal Expansion in Semiconductor Devices, IEEE Trans. Electron Devices ED-15, p. 617 (1968).

M-39. Villella, F., and Nowakowski, M. F., Investigation of Fatigue Problem in 1-mil Diameter Thermocompression and Ultrasonic Bonding of Aluminum Wire, NASA Technical Memorandum, NASA TM-X-64566 (1970). Also, Nowakowski, M. F. and Villella, F., Thermal Excursion Can Cause Bond Problems, 9th Annual Proc. IEEE Reliability Physics Symposium, Las Vegas, Nevada, pp. 172-177 (1971).

M-40. For a summary and recommendations from Reference [M-39], see Villella, F., and Martin, R., Does Your Bonding Process Doom Devices to Failure?, Circuits Manufacturing, pp. 22-30, January 1973.

M-41. Ravi, K. V., and Philofsky, E. M., Reliability Improvement of Wire Bonds Subjected to Fatigue Stresses, 10th Annual Proc. IEEE

Reliability Physics Symposium, Las Vegas, Nevada, pp. 143-149 (1972).

M-42. Phillips, W. E., in Microelectronic Ultrasonic Bonding, G. G. Harman, Ed., Nat. Bur. Stands. (U.S.), Spec. Publ. 400-2, pp. 80-86 (1974).

M-43. Riches, J. W., Sherby, O. D., and Dorn, J. E., The Fatigue Properties of Some Binary Alpha Solid Solutions of Aluminum, Trans. ASM 44, pp. 882-895 (1952).

M-44. Kinsman, K. R., Natarajan, B., and Gealer, C.A., Coatings for Strain Compliance in Plastic Packages: Opportunities and Realities, Thin Solid Films 166, pp. 83-96 (1988).

M-45. Shirley, C. G., and Blish, R. C., Thin-Film Cracking and Wire Ball Shear in Plastic DIPS due to Temperature Cycle and Thermal Shock, 25th Annual Proc., Reliability Physics Symposium, San Diego, California, April 7-9, 1987, pp. 238-249.

INDEX

By Subject

By Author